9th International Workshop on Processing and Applications of Superconducting (RE)BCO Large Grain Materials (PASREG2015)

Journal of Physics: Conference Series Volume 695

Liege, Belgium
2 – 4 September 2015

Editors:

Philippe Vanderbemden
Benoit Vanderheyden
Herbert Freyhardt
David Cardwell

ISBN: 978-1-5108-2237-5
ISSN: 1742-6588

Printed from e-media with permission by:

Curran Associates, Inc.
57 Morehouse Lane
Red Hook, NY 12571

CURRAN ASSOCIATES INC.
proceedings
.com

Some format issues inherent in the e-media version may also appear in this print version.

Printed by Curran Associates, Inc. (2016)

For permission requests, please contact the Institute of Physics
at the address below.

Institute of Physics
Dirac House, Temple Back
Bristol BS1 6BE UK

Phone: 44 1 17 929 7481
Fax: 44 1 17 920 0979

techtracking@iop.org

Additional copies of this publication are available from:

Curran Associates, Inc.
57 Morehouse Lane
Red Hook, NY 12571 USA
Phone: 845-758-0400
Fax: 845-758-2633
Email: curran@proceedings.com
Web: www.proceedings.com

Table of contents

Volume 695

9th International Workshop on Processing and Applications of Superconducting (RE)BCO Large Grain Materials (PASREG2015)
2–4 September 2015, Liege, Belgium

Accepted papers received: 23 February 2016
Published online: 17 March 2016

Preface

A related focus issue on processing and applications of (RE)BCO and MgB2 bulk superconductors is published in Superconductor Science and Technology.

Papers

PASREG 2015 (9th International Workshop on Processing and Applications of Superconducting (RE)BCO Large Grain Materials) Editorial section

The 9th International Workshop on Processing and Applications of Superconducting (RE)BCO Large Grain Materials (PASREG 2015) was held at the University of Liege, Belgium on September 02 - 04, 2015, as a satellite conference of the European Conference on Applied Superconductivity (EUCAS 2015). The Chairman was Philippe Vanderbemden; Benoît Vanderheyden, David Cardwell, and Herbert C. Freyhardt served as Co-Chairmen. PASREG 2015 was the ninth in a series of international workshops previously held in Cambridge, UK (1997), Morioka, Japan (1999), Seattle, USA (2001), Jena, Germany (2003), Tokyo, Japan (2005), Cambridge, UK (2007), Washington D.C., USA (2010), and Tainan, Taiwan (2012).

PASREG 2015 covered novel and important processing aspects of bulk High Temperature Superconductors (HTS) as well as bulk MgB_2. Besides processing, the workshop highlighted advanced material characterization technologies and activation techniques, which revealed the superior superconducting properties of bulk HTS and MgB_2. Novel developments and innovative applications involving bulk superconductors complemented the discussion and exchange of information. The workshop was attended by 63 participants from 19 different countries. A total of 50 contributions were presented, of which 31 were presented in oral form and 19 were presented as posters.

Together with a 'Special Issue' of *Superconductor Science and Technology*, these proceedings gather a selection of original research papers presented at the workshop (including both oral and poster presentations). We greatly acknowledge all the effort of the authors.

Philippe Vanderbemden
Benoît Vanderheyden
Herbert Freyhardt
David Cardwell

Journal of Physics: Conference Series **695** (2016) 011002 doi:10.1088/1742-6596/695/1/011002

Peer review statement

All papers published in this volume of *Journal of Physics: Conference Series* have been peer reviewed through processes administered by the proceedings Editors. Reviews were conducted by expert referees to the professional and scientific standards expected of a proceedings journal published by IOP Publishing.

Pinning and trapped field in MgB_2- and MT-YBaCuO bulk superconductors manufactured under pressure

T Prikhna[1,12], M Eisterer[2], X Chaud[3], H W Weber[2], T Habisreuther[4], V Moshchil[1], A Kozyrev[1], A Shapovalov[1], W Gawalek[5], M Wu [5], D Litzkendorf[4], W Goldacker[6], V Sokolovsky[7], V Shaternik [8], J Rabier[9], A Joulain[9], G Grechnev`[10], V Boutko[11] , A Gusev[11], A Shaternik[1], P Barvitskiy[1]

[1] V. Bakul Institute for Superhard Materials of the National Academy of Sciences of Ukraine, 2, Avtozavodskaya Str., Kiev 07074, Ukraine

[2] Atominstitut, Vienna University of Technology, Stadionallee 2, 1020 Vienna, Austria

[3] Laboratoire National des Champs Magnétiques Intenses (LNCMI/CNRS), 25 Avenue des Martyrs, BP 166, 38042 Grenoble Cedex 9, France

[4] Leibniz-Institut für Photonische Technologien, Albert-Einstein-Straße 9, 07745 Jena, Germany

[5] Magnetworld AG, Buchaer Straße 6, 07745 Jena, Germany

[6] Karlsruhe Institute of Technology (KIT), 76344 Eggenstein, Germany

[7] Ben-Gurion University of the Negev, P.O.B. 653, Beer-Sheva 84105, Israel

[8] G.V. Kurdyumov Institute for Metal Physics of the National Academy of Sciences of Ukraine, 03680 Kyiv, Ukraine

[9] Institute P', PHYMAT, UMR 6630, CNRS Université de Poitiers, SP2MI, BP 30179, F-86 962 Futuroscope, France

[10] B. Verkin Institute for Low Temperature Physics of the National Academy of Sciences of Ukraine, Kharkov 61103, Ukraine

[11] Donetsk Institute for Physics and Engineering named after O.O. Galkin of the National Academy of Sciences of Ukraine, R. Luxemburg str.72, Donetsk-114, 83114, Ukraine

E-mail: [1] prikhna@mail.ru, [2] eisterer@ati.ac.at, [3] tobias.habisreuther@ipht-jena.de, [4] xavier.chaud@lncmi.cnrs.fr, [5] wuminzhi@aol.com, [6]wilfried.goldacker@kit.edu, [7] sokolovv@bgu.ac.il, [8]shaternik@mail.ru, δ[9]jacques.rabier@univ-poitiers.fr, [10] grechnev@ilt.kharkov.ua, [11]boutko@teor.fti.ac.donetsk.ua

Abstract. The relevant pinning centers of Abrikosov vortices in MgB_2–based materials are oxygen-enriched Mg-B-O inclusions or nanolayers and inclusions of MgB_x (x>4) phases. The high critical current densities, j_c, of 10^6 and $10^3 A/cm^2$ at 1 and 8.5 T, respectively, at 20 K can be achieved in polycrystalline materials (prepared at 2 GPa) containing a large amount of admixed oxygen. Besides, oxygen can be incorporated into the MgB_2 structure in small amounts ($MgB_{1.5}O_{0.5}$), which is supported by Auger studies and calculations of the DOS and the binding energy. The j_c of melt textured $YBa_2Cu_3O_{7-\delta}$ (or Y123)-based superconductors (MT-YBaCuO) depends not only on the perfectness of texture and the amount of oxygen in the

[12] To whom any correspondence should be addressed.

Y123 structure, but also on the density of twins and micro-cracks formed during the oxygenation (due to shrinking of the *c*-lattice parameter). The density of twins and micro-cracks increases with the reduction of the distance between Y_2BaCuO_5 (Y211) inclusions in Y123. At 77 K j_c=8·10^4 A/cm^2 in self-field and j_c=10^3 A/cm^2 at 10 T were found in materials oxygenated at 16 MPa for 3 days with a density of twins of 22–35 per μm (thickness of the lamellae: 45-30 nm) and a density of micro-cracks of 200–280 per mm. Pinning can occur at the points of intersection between the Y123 twin planes and the Y211 inclusions. MT-YBaCuO at 77 K can trap 1.4 T (38×38×17 mm, oxygenated at 0.1 MPa for 20 days) and 0.8 T (16 mm in diameter and 10 mm thick with 0.45 mm holes oxygenated at 10 MPa for 53 h). The sensitivity of MgB_2 to magnetic field variations (flux jumps) complicates estimates of the trapped field. At 20 K 1.8 T was found for a block of 30 mm in diameter and a thickness of 7.5 mm and 1.5 T (if the magnetic field was increased at a rate of 0.1 T) for a ring with dimensions 24×18 mm and a thickness of 8 mm.

1. Introduction

Both MgB_2- and MT-YBaCuO bulk materials are good candidates for the same cryogenic applications. The working temperature of MT-YBaCuO can be higher (77 K) than that of MgB_2 (20-30 K), while MgB_2-based materials are much cheaper and easier to prepare. Excellent superconductive and mechanical characteristics and well developed manufacturing technologies make these materials extremely promising for fault current limiters (FCL), electrical machines, for screening magnetic and gravitation fields, for magnetic bearings, contactless mixers, as magnets for magnetron sputtering, for MAGLEV transportation (particularly in clean rooms) and other devices based on the principle of levitation, etc.

The manufacturing technologies for the preparation of superconducting materials with high critical current densities, j_c, are very different for MgB_2 and MT-YBaCuO due to the difference of their coherence lengths. The comparatively large coherence length of MgB_2 allows attaining high critical currents in polycrystalline materials, because the grain boundaries contribute to pinning. Therefore, the synthesis technologies aim at the formation of dense nanostructural MgB_2–based materials with long lengths of grain boundaries. Hot pressure (at 30 MPa), spark plasma or SPS (at 50 MPa) and high pressure (2 GPa) synthesis allow attaining j_c of around 4·10^5 - 10^6 A/cm^2 at 20 K in a 1 T field [1, 2] and high trapped fields [3].

In polycrystalline $YBa_2Cu_3O_{7-\delta}$ (Y123) the grain boundaries (even the misalignment of grains by more than 5 degrees) are obstacles for the superconducting current flow because of the small coherence length of the compound, which makes it impossible to attain j_c above 10^3 A/cm^2 at 77 K in self-field. Therefore, technologies of melt textured growth on seed crystals were developed (i.e. the formation of a pseudo-single crystalline Y123 structure with a high density of small non superconducting inclusions of the Y_2BaCuO_5 (Y211) phase), which allow attaining j_c in self-field above 6·10^4 A/cm^2 in MT-YBaCuO at 77 K and trapped fields of around 1 T (depending on the size of the pellets) [4-7].

Because of the difference in coherence length, the size of the inhomogeneities which can act as pinning centers in bulk MgB_2-based materials and in MT-YBaCuO, are different (they should be comparable to the coherence length). Thus in Y123 twins, dislocations, stacking faults, and possibly micro-cracks (formed because of shrinking of *c*-parameter of Y123 during oxygenation) [4, 8-10] can be responsible for pinning, whereas in MgB_2 grain boundaries of nano-grains of the matrix phase (with a stoichiometry close to MgB_2), nanosized inclusions of MgO, Mg-B-O and higher magnesium borides MgB_x($x \geq 4$) or inclusions of other secondary phases can act as pinning centers [11-13]. In MT-NEG123 (or Nd-Eu-Gd-Ba-Cu-O-based melt-textured materials) an array of nanolamellas within regular twins has been proposed as pinning centers responsible for very high irreversibility fields [14]. The distance between the nanolamellas was only several nanometers. The nanolamellas inside the twins were not observed when the amount of NEG211 in MT-NEG123 increased above 10 mol.%, which was accompanied with the reduction of the irreversibility field.

It is well known that for attaining high transition temperatures and critical current densities in $YBa_2Cu_3O_{7-\delta}$ the concentration of oxygen (7-δ) in the unit cell and its distribution in the basal planes are very important (the formation of oxygen chains in the basal planes and an amount of oxygen 7-δ~6.9-7.0 are optimal) [8, 15]. For the preparation of MT-YBaCuO with good superconducting properties the oxygenation stage (usually a separate process) is essential, which is needed in order to transform the non superconducting $YBa_2Cu_3O_{7-\delta}$ matrix phase of MT-YBaCuO with 7-δ~6.3-6.4 into the superconducting phase with 7-δ~6.9-7.0. During the oxygenation process twins, dislocations, stacking faults, micro- and macro-cracks are formed, because the oxygen incorporation into the $YBa_2Cu_3O_{7-\delta}$ structure is accompanied by a shrinking *c*-axis lattice parameter.

Although MgB_2 is nominally an oxygen free compound, admixed oxygen is usually present, even if high purity initial ingredients are used and the preparation is conducted under protective atmosphere.

Our results support the conclusion that excellent SC properties (critical current density and thus trapped magnetic field) of MT-YBaCuO are connected to a high extent with a high density of twins, which can be influenced by pressure-temperature conditions during the oxygenation process and by the density and homogeneous distribution of Y211 inclusions formed during melt-texturing (which, in turn, even depend on the kind of initial ingredients). The superconducting performance of MgB_2-based materials essentially depends on the distribution of admixed oxygen in the form of Mg-B-O nanoinclusions or nanolayers (with most probably MgO structure) and on the amount of oxygen dissolved in the hexagonal superconducting matrix, which can be regulated by the temperature – pressure conditions during manufacturing and by additions, such as Ti and SiC, for example. The solution of some oxygen in the superconducting MgB_2 structure does not contradict its SC performance according to the calculated density of states and the binding energy. Furthermore, it was shown that the level of trapped fields in bulk MgB_2 depends to a high extent on the penetration rate of the external magnetic field into the material, because too high rates provoke flux jumps leading to a destruction of the trapped magnetic field.

2. Experimental

In this study, three types of MT–YBCO initial ceramic samples were used: (type 1) - the traditional bulk MT–YBaCuO ceramic prepared according to the technology described in [6]. As the initial material a mixture of powdered $YBa_2Cu_3O_{7-\delta}$ (produced by Solvay Company), Y_2O_3, and CeO_2 in the ratio of $Y_{1.5}Ba_2Cu_3O_{7-\delta}$ + 1% CeO_2 was used. The powders were thoroughly mixed and uniaxially compacted as square blocks with dimensions 40×40×20 mm^3. Then a $SmBa_2Cu_3O_x$ seed crystal was placed onto the upper surface of each block and a modified process of melt texturing of a batch of 16–30 samples was performed under quasi-isothermal conditions in a furnace with six-sector heating capability. After melt-texturing, due to shrinking and after cutting off the bottom, the sample dimensions were 38×38×17 mm^3. Texturing and cooling after texturing were carried out in air. Melt-textured ceramics of type 2 were prepared without and with small parallel holes (to reduce the depth into which oxygen could penetrate during oxygenation). The thin-wall MT-YBaCuO ceramics with holes do not contain pores and macro-cracks. MT-YBaCuO materials of type 2 (thin-walled and traditional) were prepared as described in [16] using $SmBa_2Cu_3O_x$ seed crystals (after finishing the texturing process air was slowly substituted by nitrogen; the samples were cooled from about 940 °C in nitrogen with a residual oxygen pressure below 5 mbar to avoid oxygen uptake and limit cracking due to oxygenation). Bulks with a diameter of about 20 mm with small parallel holes about 0.7 and 0.5 mm in diameter and with 1.2 or 2.5 mm spacing between each other were formed from the initial mixture, which contained 70 $YBa_2Cu_3O_{7-\delta}$, 30 Y_2BaCuO_5, and 0.15 PtO_2 wt %. After melt-texturing, the pellet had a diameter of 16 mm and the holes had a reduced diameter of about 0.6 mm and 0.45 mm. The top and bottom parts were cut and polished. The final thickness was about 10 mm. The holes did not prevent the growth of a single-domain structure from the seed crystal surface within the whole volume of a sample.

In this study, the initial ceramic blocks of types (1 and 2) were cut into rectangular parallelepipeds

Table1. Characteristics of initial boron powders

Type of amorphous boron (producer)	N ; wt %	C, wt %	H, wt %	O, wt %	average grain size
B(I) (H.C. Starck)	0.48	0.31	0.32	0.66	<5 μm
B(II) (HyperTech, USA)	1.02	3.5	0.87	-	<1 μm
B(III) (H.C. Starck)	0.40	0.47	0.37	1.5	4.0 μm
B(IV) (MaTeck 95-97%)	0.43	0.27	0.11	1.9	1.4 μm

measuring 3–4×3–4×5–8 mm in such way that the *c* crystallographic axis of Y123 was parallel to the longer side of parallelepipeds and the *ab*–plane parallel to the shorter dimensions. These sizes are suitable for placing the samples into a vibrating sample magnetometer for j_c measurements, which were performed in two perpendicular directions. The small samples were oxygenated together with the big ones, which were used for the trapped field measurements. The phase composition and the crystallographic structures were studied at room temperature by x-ray powder diffraction using a Philips X'pert" and a DRON-3 diffractometer. The x-ray patterns were taken in an angular range of 2Θ = 20–70° at a rate of 5 deg/min. The sample structure was studied using polarized light and transmission electron microscopes.

The bulk MgB_2 materials were synthesized from Mg and B taken in Mg:2B stoichiometry at 2 GPa quasi-hydrostatic pressure and at 600 - 1050 °C for 1 h (in contact with BN). We used Mg turnings Technical Specification of Ukraine 48-10-93-88 (designated as Mg(I)) or fine Mg powder produced by HyperTech, USA (designated as Mg(II)) and commercially available amorphous precursor B powders (Table 1). B and Mg were milled in a high-speed planetary activator with steel balls for 1-3 min.

Thin MgB_2 films were deposited by DC magnetron sputtering under 1 Pa Ar pressure on 8×8×0.2 mm^3 sized sapphire substrates in (0001) orientation at room temperature and subsequently annealed at 600-650 °C for 5 min in Ar under 10 Pa. For deposition we used an MgB_2 target synthesized by hot pressing at 30 MPa, 800 °C, 1 h using boron B(II).

The microstructure of the materials was characterized by x-ray structure analysis (with Rietveld refinement) and by SEM with microprobe x-ray and Auger (JAMP−9500F) analysis. The Auger spectroscope JAMP−9500F is combined in the same device with a high resolution SEM. It is possible to change the SEM microprobe analyzer for the Auger probe and to perform etching by Ar ions of the 1×1 mm^2 analyzed material surface area. JAMP−9500F offers the possibility to quantitatively analyze small volumes of about 10 nm in diameter and 1 nm in depth. Thus, non-oxidized MgB_2 surfaces can be analyzed as well as the composition of nano-inclusions and nano-layers.

The critical current density, j_c, was estimated from magnetization measurements in an Oxford Instruments 3001 vibrating sample magnetometer (VSM) using the Bean model. The connectivity was estimated from the difference in resistivity at 40 K and 300 K measured by the four probe technique. The amount of the SC shielding fraction was calculated from the ac susceptibility at 5 K with a numerical correction for the demagnetization of the actual sample geometry.

3. Results and discussion

3.1 Structure, pinning and critical currents in MT-YBaCuO

Figure 1 summarizes the results on MT-YBaCuO. The structures of type 1 and 2 MT-YBaCuO are shown in figures 1a-d as obtained from optical microscopy in polarized light (polished and etched surfaces) after oxygenation at 0.1 MPa (1 atm) pressure of oxygen at 440 °C for 20 days (figures 1a, b) and after oxygenation at 16 MPa oxygen pressure, 800 °C for 3 days (figures 1c, d). Figures 1e, f, g show the whole samples after oxygenation and polishing as they look under optical microscopy in polarized light and ware composed from dozens of pictures taken one by one. Figures 1h-k present the structures obtained using transmission electron microscopy of type 1 and 2 MT-YBaCuO oxygenated under 16 MPa pressure at 650 and 800 °C. Figure 2 depict the j_c of MT-YBaCuO oxygenated under

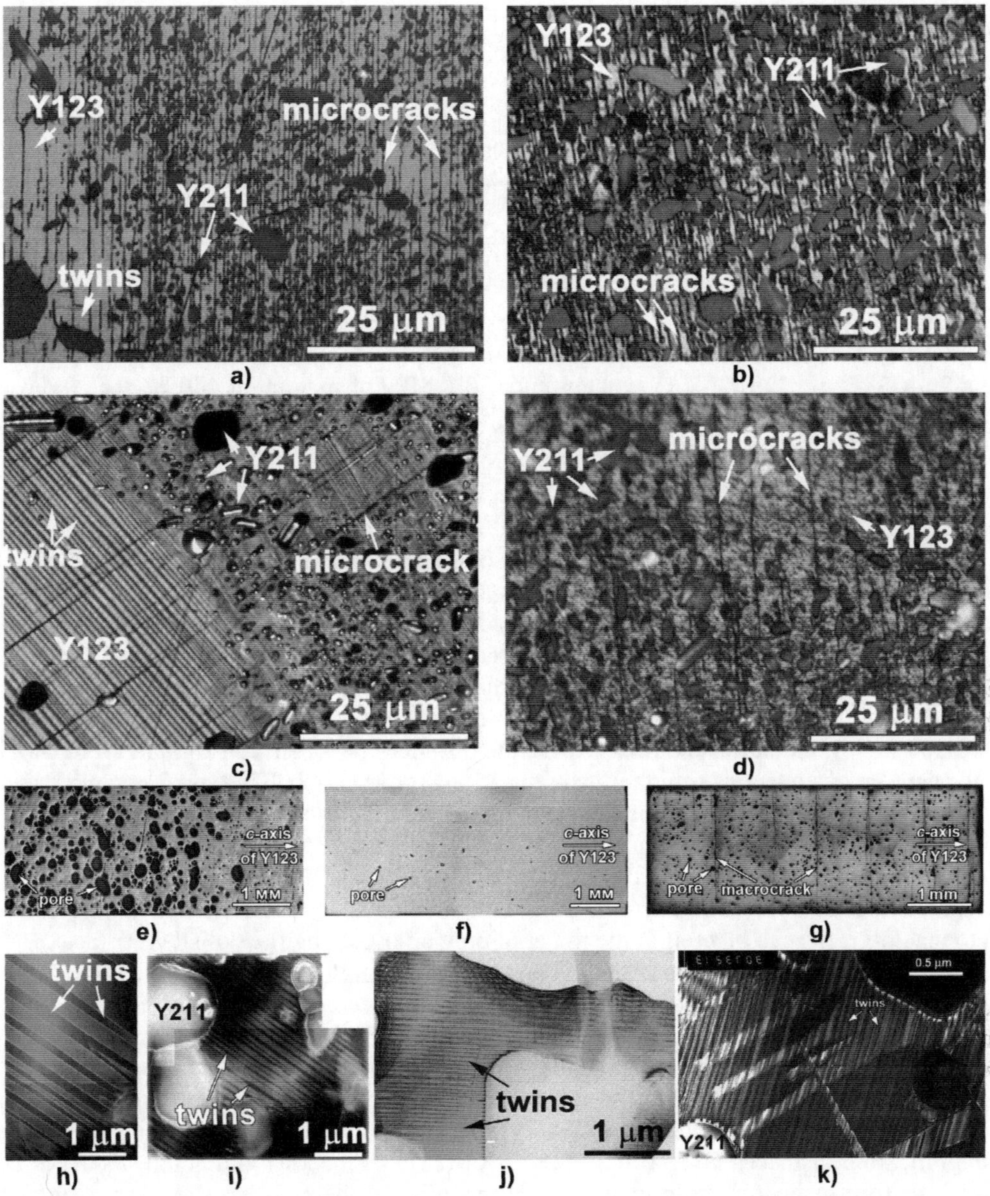

Figure 1. (a-d) – Structures of MT-YBaCuO ceramics under polarized light after oxygenation and etching: of type 1 (a, c) and type 2 (b, d), oxygenated under 0.1 MPa of oxygen at 440 °C for 20 days (a, b) and under 1–16 MPa of oxygen at 900–800 °C for 3 days (c, d); (e-k) view of the structure of MT-YBaCuO samples: (e-g) under polarized light as a whole which were used for j_c estimation (the longer side of the rectangular parallelepiped sample was parallel to the *c*-axis of Y123) and (h-k) under TEM of: (e, h, i) type 1 oxygenated under 1–16 MPa of oxygen at 900–800 °C for 3 days, (f, j) type 2 (with 0.6 mm holes) oxygenated under 1–16 MPa of oxygen at 900–800 °C for 3 days and (g, k) type 2 (traditional without holes) oxygenated under 16 MPa of oxygen at 700 °C for 3 days.

different conditions in two perpendicular directions: parallel (figure 2a, b) and perpendicular (figure 2c) to the *ab* planes. It should be mentioned that the lattice parameters of the $YBa_2Cu_3O_{7-\delta}$ phase of all MT-YBaCuO samples after oxygenation corresponded to an amount of oxygen of about 7 ($\delta \approx 0$). This means that the Y123 structure was completely oxygenated. The analysis of these results allows us to

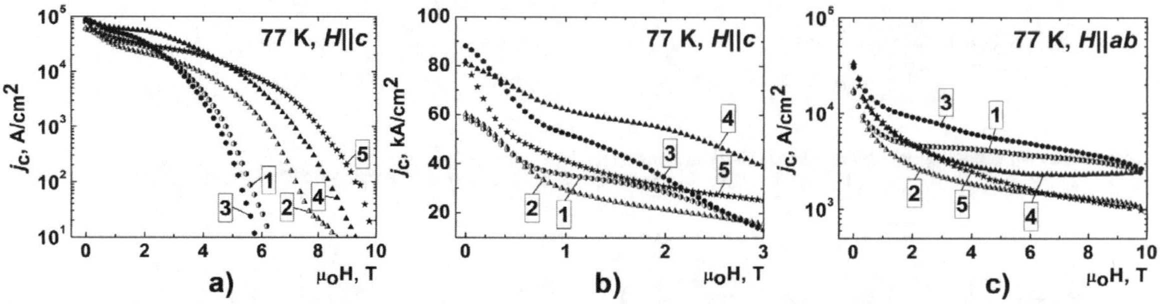

Figure 2. Critical current density, j_c, vs. magnetic field, $\mu_o H$, at 77 K when the magnetic field was parallel to the *c*-axis of Y123 (a, b) and parallel to the *ab*-planes (c) of the MT-YBCO samples: 1, 3 – type 1 and 2, 4, 5 – type 2: 1, 2 - oxygenated under 0.1 MPa of oxygen at 440 °C for 20 days; 3, 4 – oxygenated under 1-16 MPa of oxygen at 900-800 °C for 3 days and 5 - oxygenated under 16 MPa of oxygen at 700 °C for 3 days; 1, 3, 5- traditional MT-YBaCuO without holes and 2, 4 – thin-walled MT-YBaCuO with 0.6 mm holes.

draw the following conclusions. Samples oxygenated under high oxygen pressure (16 MPa) exhibit higher critical current densities, especially in the case of type 2 MT-YBaCuO prepared from an Y123 - Y211 mixture. Comparing the structures in figures 1a-d the following differences are found. First of all, ceramics of type 1 (prepared from Y123 and Y_2O_3) have smaller and less homogeneously distributed inclusions of the Y211 phase as compared to ceramics of type 2 (prepared from Y123 and Y211). For both ceramics an essential decrease of the amount of micro-cracks parallel to the *ab*-plane was observed in the ceramic oxygenated under high oxygen pressure (factor of 4 - 4.5, i.e. from 1080-980 to 280-200 mm^{-1}), but at those places, where the distances between Y211 grains were smaller, the density of micro-cracks was much higher. The distances between Y211 in type 2 ceramics were very small and they were more homogeneously distributed, so the micro-crack density is higher than that in type 1 ceramics (factor of 1.4 for oxygenation at 16 MPa pressure). This can explain why the critical current densities of type 1 ceramics are somewhat higher (curves 1 and 3 vs. curves 2 and 4) in the direction perpendicular to the *ab*-plane (figure 2c). It should be mentioned that micro-cracks as opposed to macro-cracks can be seen in the material only after etching the polished surfaces.

The analysis of the TEM images shows the extremely high twin density in type 2 ceramics oxygenated at 650 and 800 °C under 16 MPa (figures 1j, k) and a somewhat lower density in type 1 ceramics oxygenated at 800 °C under 16 MPa (see figures 1 h, i). The twin density depends on the distribution of Y211 inclusions in Y123 as well and was higher when the distance between the Y211 inclusions was smaller. In type 1 ceramics twins can be seen even in optical microscopy in polarized light (figure 1c) at those places where the density of Y211 is low, but twins can be seen only under higher magnification at places with a high density of Y211 inclusions and small distances between them, which can be reached in transmission electron microscopes (TEM), because their density is extremely high and the twin lamella thickness is around 30-45 nm. The density of twins in high pressure oxygenated MT-YBaCuO of type 2 was about 22-35 μm^{-1} (figure 1j, k), while the ceramics of type 1 had a twin density of 4-11 μm^{-1} (in those regions, where the Y211 density was lower, a lower twin density was observed, see figures 1h, i). The densities of twins in materials oxygenated at 440 °C and 0.1 MPa were around 0.5-1.7 and 12-16 μm^{-1} for MT-YBaCuO of types 1 and 2, respectively.

It is interesting to note that high pressure oxygenation at 800 °C results in MT-YBaCuO materials, where almost only twins are present in the Y123 matrix, but no stacking faults and only very randomly located dislocations. According to the diagram of equilibrium concentration of oxygen in the Y123 structure versus the partial oxygen pressure at different temperatures by Assmus and Schmidbauer [15], full oxygenation of Y123 (7 oxygen atoms per unit cell) under 10-16 MPa pressure at such high temperature as 800 °C is impossible and 5-8 times higher pressures are necessary to keep the oxygen in Y123. But it was experimentally shown that the diffusion processes are rather quick at such temperatures and oxygen can penetrate into the Y123 structure easily, provoking just twinning and much less cracking than in the case of lower oxygenation temperatures. An analysis of the MT-YBaCuO structure, high pressure oxygenated at 650 °C (figures 1g, k), showed the presence of a lot of macro-cracks (figure 1g), 500 mm^{-1}, a very high concentration of twins (30-35 μm^{-1}) as well as stacking faults and dislocations. The high density of cracks can be an explanation for the comparatively low j_c in the direction perpendicular to the *ab*-planes (curve 5 in figure 2c). So, the temperature of 650 °C was not high enough to oxygenate the material without cracking. It should be mentioned that even small samples could not be oxygenated at 440 °C at least for 165 h, when an oxygen pressure of 16 MPa was applied before heating, i.e. the Y123 structure remained tetragonal and zero j_c was observed at 77 K. The oxygenation of small samples at 440 °C in flowing oxygen (0.1 MPa) for 3 days gave much better results from the point of view of j_c, but it was still much lower than that of both types of samples oxygenated at 800 °C and 16 MPa for the same time and their Y123 structures were partly tetragonal and partly orthorhombic.

The MT-YBaCuO materials with small holes (thin-walled ceramics) and without holes (traditional ceramics) have a different porosity (their structures are shown in figures 1f and 1g, respectively). Usually a dense layer is formed on the surfaces of traditional ceramics, while highly porous material is located below, but the thin-walled ceramics are completely dense. We studied in parallel traditional porous MT-YBaCuO ceramics prepared from a mixture of Y123 and Y211 (type 2) and conclude that porous ceramics of type 1 and type 2 behave differently concerning the amount of micro-cracks and twins because of the different distribution of Y211 particles, but not because of a difference in porosity, because the type 2 porous ceramics behave similarly to type 2 ceramics with holes. The high pressure-high temperature oxygenation (800 °C, 16 MPa) allowed avoiding macro-cracking both in porous and dense MT-YBaCuO, prepared from Y123 and Y_2O_3 (type 1) and from Y123 and Y211 (type 2), see figures 1e, f.

Our experimental study showed that high critical currents were achieved in MT-YBaCuO with a high density of twins (and without dislocations and stacking faults). The authors of [4] observed an essential increase of j_c in MT-YBaCuO at low magnetic fields (below 0.3 T) after repeated oxygenation at 10 MPa and 600 °C for 2-12 h of ceramics oxygenated previously at 450 °C under 0.1 MPa for 120 h, but the j_c in fields above 0.3 T decreased. The increase of j_c was attributed by the authors of [4] only to the increase of dislocation density, but not to the density of twins (which increased by a factor of 1.7 as well, as our observations of the structures demonstrated). Our previous study of MT-YBaCuO samples, detwinned under high pressure (2 GPa) and having an extremely high density of dislocations (10^{12} cm^{-2}), exhibited a rather modest j_c (about 10^3 A/cm^2 in self-field). Taking all observations into consideration we come to the conclusion that in addition to the intersection places between twins pinning will occur at those places, where the twin planes cross 211 inclusions or micro-cracks and that twins play a very important role for attaining high critical currents. Usually an increase of micro-hardness and fracture toughness [17] is observed in materials oxygenated under high pressure - high temperature, which could well be the result of a decrease of the micro-crack density and an increase of the density of twins.

3.2 Structure, pinning and critical currents in MgB_2

Figure 3 summarizes our results on MgB_2 bulks and thin films. The highest j_c showed a thin MgB_2 film (curves 1, 2 in figures 3g, h), which contained a lot of admixed oxygen in its structure according to SEM and Auger studies. Oxygen is not uniformly distributed over the film surface (the brighter areas in figures 3e, f correspond to higher concentrations of oxygen, the surface of the film was etched by Ar ions in the chamber of the JAMP−9500F instrument before the images were taken). X-ray examinations confirmed the MgB_2 structure of the film. For bulk MgB_2 materials j_c usually increases with increasing manufacturing temperature at low magnetic fields, while j_c somewhat decreases at

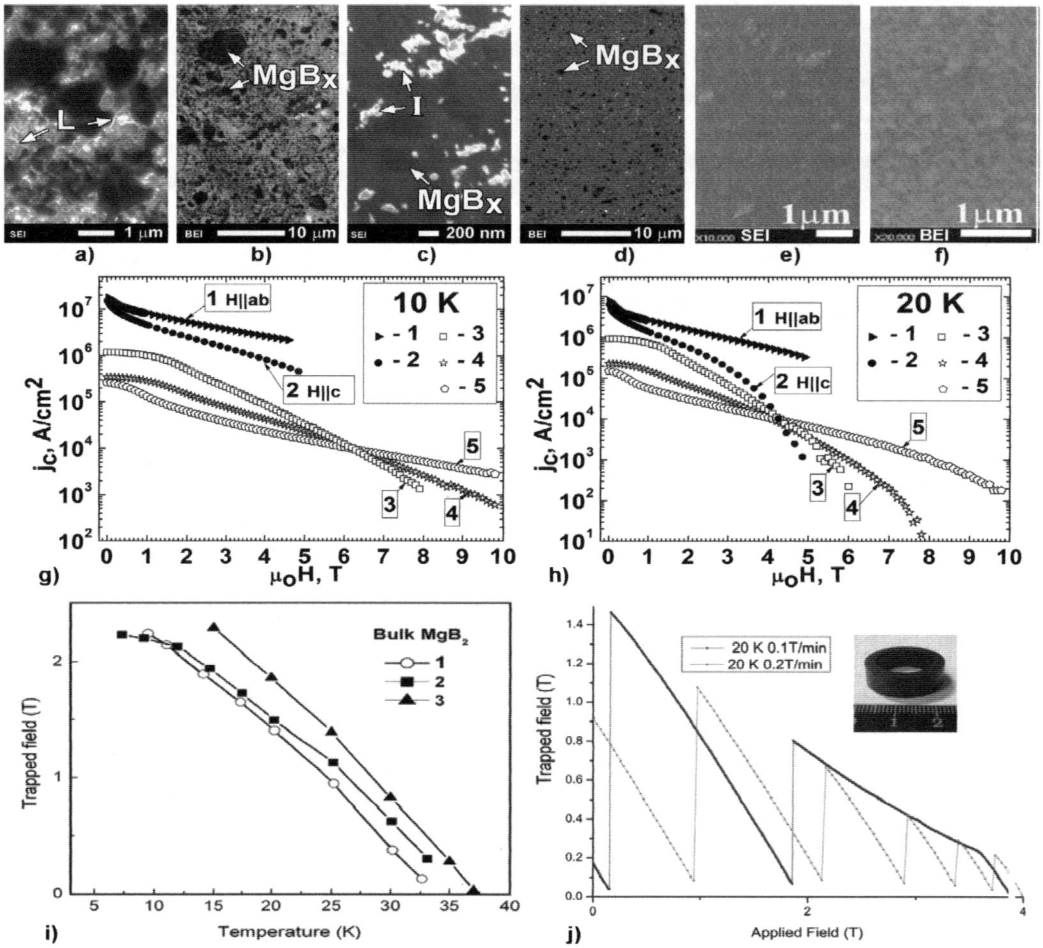

Figure 3. (a–d) - Structures of MgB_2 bulk (after etching by argon in the chamber of JAMP−9500F) synthesized from Mg(I):2B(I) under 2 GPa for 1 h at 800 °C (a, b) and at 1050 °C (c, d) in SEI (a, c) and BEI (b, d) modes ("L", "I" – Mg-B-O nano-layers and inclusions, respectively, MgB_x – inclusions of higher magnesium borides x>4);
(e-f) surface of the MgB_2 film (140 nm thick, grown by magnetron sputtering on sapphire substrate, after etching by argon in the chamber of JAMP−9500F) in SEI (e) and BEI (f) modes (the places with higher concentration of admixed oxygen look brighter);
(g, h) – critical current density, j_c, vs. magnetic field, $\mu_o H$, at 10 K (g) and 20 K (h) for the thin film (1, 2): (magnetic field parallel to the substrate – 1 and perpendicular – 2) and for bulks (3–5), high-pressure (2 GPa, 1 h) MgB_2 synthesized at: 3 - 1050 °C from Mg(I):2B(I), 4 - 800 °C from Mg(I):2B(I) and 5 - 600 °C from Mg(I):2B(II);
(i) - trapped fields at the sample centre of MgB_2 blocks of different sizes: 1 - 14×14×4 mm^3 (from Mg(I):2B(IV) with 10% Ti), 2 - 22.5 mm in diameter and 6 mm in height (from Mg(I):2B(IV) with 5% of Ta), 3 - 30 mm in diameter and 7.5 mm in height (from Mg(I):2B(III) with 10% Ti);
(j) trapped field at 20 K at the center of the MgB_2 ring from Mg(I):2B(III) with 10% SiC (Ø24.3×17.9 mm, h=7.8 mm, see upper right insert). The measurements were performed with different sweep rates of the external magnetic field (0.1 T/min: solid line, 0.2 T/min: gray line with squares).

high magnetic fields (curves 3-5 in figures 3g, h), which is accompanied by a transformation of Mg-B-O nanolayers into separate Mg-B-O inclusions (the brightest areas in figures 3a, c) and by a reduction of MgB_x (x >4 and usually x=10-14 for materials prepared at 2 GPa; darkest areas in figures 3d, c).

Comparing the structures and characteristics of MgB_2 we conclude that the most important role for pinning is played by the distribution of admixed oxygen in the structure and that its generally high amount (5-17 wt%) is certainly not an obstacle for very good superconducting characteristics (figures 3g - 3j). MgB_2 bulks manufactured under 30 MPa – 2 GPa pressure showed 73-98 % connectivity and 75-100 % shielding fraction. We did not observe MgO by SEM and Auger techniques. The presence of some (5-13 wt%) MgO in the x-ray diffraction pattern makes us assume that the brightest Mg-B-O areas represent solid solutions of boron in the MgO structure (up to about MgBO level). Our careful Auger study [14] showed that some oxygen is solved in the MgB_2 matrix as well ($MgB_{2.2-1.7}O_{0.4-0.6}$ according to a quantitative Auger analysis). Calculations of the density of electronic states (DOS) and of the binding energy, E_b, (Table 2) confirmed that, from the point of view of carrier concentration near the Fermi level (metallic-like behaviour) and the similarity of the binding energy with that of MgB_2, the excellent superconducting performance of the $MgB_{1.5}O_{0.5}$ compound does not contradict theoretical predictions. To calculate the DOS and E_b we applied the density functional theory [18] based on a full-potential linearized augmented plane wave method with the generalized gradient correction to the exchange-correlation potential [19] using WIEN2k [20].

Table 2. Total DOSs and binding energies, E_b, in the $Mg(B_{1-x}O_x)_2$ compounds.

x	0	0.125	0.25	0.5
total DOS	0.72	0.43	0.96	1.36
E_b, Ry	1.12	1.10	1.06	0.94

3.3 Trapped fields of MT-YBaCuO and MgB_2 bulk

The trapped magnetic field is one of the most comprehensive characteristics of superconducting ceramics from the point of view of perspectives for their application. The strength of the magnetic field of a magnetized bulk depends on the material's j_c and its volume. Therefore, the size, homogeneity and perfection of the bulk structure play a very important role. The trapped field distribution (field mapping) is assessed by scanning the surface using Hall probes and represents the main characterization technique for MT-YBaCuO bulks. MT-YBaCuO blocks of 38×38×17 mm^3 (type 1) oxygenated at 440 °C and 0.1 MPa for 20 days trapped a maximal field of 1.44 T at 77 K. MT-YBaCuO (type 2) with 16 mm in diameter, 10 mm thick, and with 0.6 mm holes trapped a maximal field of 0.65 T after oxygenation at 800 °C and 16 MPa for 3 days and just 0.24 T after oxygenation at 0.1 MPa and 420-390 °C for 18 days. MT-YBaCuO (type 2) with 16 mm in diameter, 10 mm thick, and with 0.45 mm holes trapped a field of 0.84 T after oxygenation at 800 °C and 10 MPa for 53 h.

In the case of MgB_2 the working temperature is rather low and the scanning devices at such low temperatures (20 K) are rather scarce. Therefore, trapped fields of MgB_2 are often estimated from only one point (in the middle of the top surface of the bulk, where the highest trapped field value is expected). The temperature dependence of trapped fields of MgB_2 is shown in figure 3i. MgB_2 is very sensitive to flux jumps and, as shown in figure 3j, the trapped field in MgB_2 depends to a high extent on the speed of the external magnetic field variation. This makes a pulsed magnetization of MgB_2 extremely difficult and restricts applications under ac current conditions. We succeeded to trap 1.8 T at 20 K (figure 3i) in the middle of an MgB_2 block of 30 mm in diameter, 7.5 mm thick.

4. Conclusions

The investigation of the oxygenation process of MT-YBCO ceramics with different sizes and distributions of Y211 particles in the Y123 structure allowed us to conclude that twins influence pinning and thus the critical current density at elevated temperatures to a high extent, but their density as well as the density of micro-cracks parallel to the *ab*-plane depend to a high extent on the distances between Y211 inclusions. Both the density of twins and of micro-cracks increases with a reduction of these distances. Oxygenation under elevated oxygen pressure (10-16 MPa) and at high temperature (800 °C) allows to increase j_c and the trapped field, which is, as we assume, due to the enhancement of the diffusion processes and an oxygenation of the Y123 structure under softer conditions in much shorter times leading to an essential increase in twin density and a reduction of micro-cracking. Pinning in MgB_2 depends to a high extent on the distribution of admixed oxygen, which can be influenced, e.g., by the synthesis temperature. The relevant pinning centers are in our opinion nano-areas of solid solutions of boron in MgO (up to the MgBO level) and inclusions of higher magnesium borides MgB_x ($x>4$), the sizes and distribution of which also depend on the synthesis conditions. The dissolved oxygen in the MgB_2 structure does not depress superconductivity in MgB_2 also from the point of view of DOS and binding energy. The level of the trapped magnetic fields depends in MgB_2 to a high extent on the rate of the external magnetic field variation.

Acknowledgement
This work was partly performed within the Austrian-Ukrainian project UA 01/2015 and within the German-Ukrainian project HICONSU (FKZ: 01DK13031) supported by German Ministry of Education and Research (BMBF). The contents and the results represent the opinion of the authors.

References

[1] Prikhna T, Eisterer M, Gawalek W et al. 2015, *Superconductors and Superconductivity. Specialized Collections* **4** 30.
[2] Prikhna T, Noudem J, Gawalek W et al. 2012, *Materials Science Forum* **721** 3.
[3] Yamamoto A, Ishihara A, Tomita M and Kishio, K 2014 *Applied Physics Letters* **105** 032601.
[4] Plain J, Puig T, Sandiumenge F, Obradors X and Rabier J 2002 *Physical Review B* **65** 104526.
[5] Takizawa T and Murakami M 2005 Critical Currents in Superconductors, Editor M. Murakami Tokyo: Nihon University College of Humanities and Sciences.
[6] Gawalek W, Habisreuther T, Zeisberger M, Litzkendorf D, Surzhenko O, Kracunovska S, Prikhna T A, Oswald B, Kovalev L K and Canders W 2004 *Supercond. Sci. Technol.* **17** 1185.
[7] Deng Z, Izumi M, Miki M, Tsuzuki K, Felder B, Liu W, Zheng J, Wang S, Wang J, Floegel-Delor U, Werfel F N 2012 *IEEE Transactions on Applied Superconductivity* **22,** 2, 6800210.
[8] Murakami M 1993 Melt processed high-temperature superconductors, Editor Murakami M World Scientific: Singapore-New Jersey – London - Hon-Kong 380.
[9] Diko P 2004 *Supercond. Sci. Technol.* **17** 45.
[10] Boyko V S, Siu-Wai Chan 2007 *Physica C* **466** 56.
[11] Liao X Z, Serquis A C, Zhu Y T, Huang J Y, Civale L, Peterson D E, Mueller F M and Xu H 2003 *Journal of Applied Physics* **93** 6208.
[12] Prikhna T A, Eisterer M, Weber H W, Gawalek W, Kovylaev V V, Karpets M V, Basyuk T V and Moshchil V E 2014 *Supercond. Sci. Technol.* **27**, No 4 044013.
[13] Birajdar B, Braccini V, Tumino A, WenzeT, Eibl O and Grasso G 2006 *Supercond. Sci. Technol.* **19** 916.
[14] Muralidhar M, Sakai N, Chikumoto N, Jirsa M, Machi T, Nishiyama M, Wu Y, Murakami M 2002, *Phys. Rev. Lett.* **89,** No 23 237001_1.
[15] Assmus W, Schmidbauer W 1993 *Supercond. Sci. Technol.* **8** 555.
[16] Chaud X, Bourgault D, Chateigner D, Dico P, Porcar L, Villaume A, Sulpice A, Tournier R *Supercond. Sci. Technol.* 2006 **19** S590.
[17] Prikhna T, Chaud X, Gawalek W, Rabier J, Savchuk Y, Joulain A, Vlasenko A, Moshchil V, Sergienko N, Dub S, Melnikov V, Litzkendorf D, Habisreuther T, Sverdun V 2007 *Physica C* **460-462** 392.
[18] Parr R G and Yang W 1989 Density-Functional Theory of Atoms and Molecules, *Oxford Univ. Press.*
[19] Perdew J P, Burke S and Ernzerhof M 1996 *Phys. Rev. Lett.* **77** 3865
[20] Blaha P, Schwarz K, Madsen G K H, Kvasnicka D, and Luitz J 2001 WIEN2K, An augmented plane wave + local orbitals program for calculating crystal properties. *Techn. Univ. Wien.*

Microstructural and magnetic analysis of a superconducting foam and comparison with IG-processed bulk samples

A Koblischka-Veneva[1], M R Koblischka[1], N Ide[2], K Inoue[2], M Muralidhar[2], T Hauet[3], and M Murakami[2]

[1]Institute of Experimental Physics, Saarland University, Campus C 6 3, D-66123 Saarbrücken, Germany.
[2]Superconducting Materials Laboratory, Department of Materials Science and Engineering, Shibaura Institute of Technology 3-7-5 Toyosu, Koto-ku, Tokyo 135-8548, Japan.
[3]Institute Jean Lamour, UMR CNRS-Université de Lorraine, Vandœuvre-lès-Nancy, France

m.koblischka@mx.uni-saarland.de

Abstract. $YBa_2Cu_3O_y$ (YBCO) foam samples show an open, porous foam structure, which may have benefits for many applications of high-T_c superconductors. As the basic material of these foams is a pseudo-single crystalline material with the directional growth initiated by a seed crystal similar to standard melt-textured samples, the achieved texture of the YBCO is a very important parameter. We analyzed the local texture and grain orientation of the individual struts forming the foam by means of atomic force microscopy and electron backscatter diffraction (EBSD). Furthermore, the magnetic properties of a foam strut are evaluated by means of SQUID measurements, from which the flux pinning forces were determined. A scaling of the pinning forces in the temperature range between 60 K and 85 K was performed. These data and the details of the microstructure are compared to IG-processed, bulk material.

1. Introduction

For applications of high-T_c superconductors, not only high critical current densities are of importance, but also the weight and the production costs play a significant role. In order to develop new types of superconducting material, Reddy *et al.* [1,2] developed superconducting foams with an open-cell structure. These materials offer a light weight, the possibility to produce very large sample sizes, as well as an easy oxygenation process. The fabrication of these foams is based on polyurethane foams, which are coated with a slurry of Y_2BaCuO_5 (211), and then the material is transformed into superconducting YBCO using an infiltration growth (IG) process. The IG process can be applied to the fabrication of large, bulk YBCO samples as well as described in Refs. [3-6]. Therefore, it is interesting to compare the superconducting properties (critical current density, flux pinning) and the achieved microstructure of the foams and the IG-processed samples with each other.

2. Experimental procedure

The foam sample studied here was fabricated at ACCESS, Aachen as described in Refs. [1,2,7]. The IG-processed YBCO samples were produced at SIT, Tokyo according to Refs. [5,6]. A foam strut was broken out of the big piece at a position close to the seed (position (1) in Ref. [7]). For the IG-processed sample, pieces of $1.5 \times 1.5 \times 0.5$ mm^3 were cut from the bulk using a diamond saw.

For the characterization of the microstructure, all sample surfaces were mechanically polished prior to scanning, either dry from 12 μm to 0.5 μm 3M diamond paper or wet from 320 grain SiO paper to 4000 grain SiO paper and then from 3 μm diamond polishing solution down to 40 nm colloidal silica suspension [8]. After that, the samples were cleaned for several minutes in acetone in an ultrasonic bath and then for several minutes in an ethanol bath. Both surface preparation methods serve well for the AFM/STM measurements, as well as for the electron backscatter diffraction (EBSD) analysis [8].

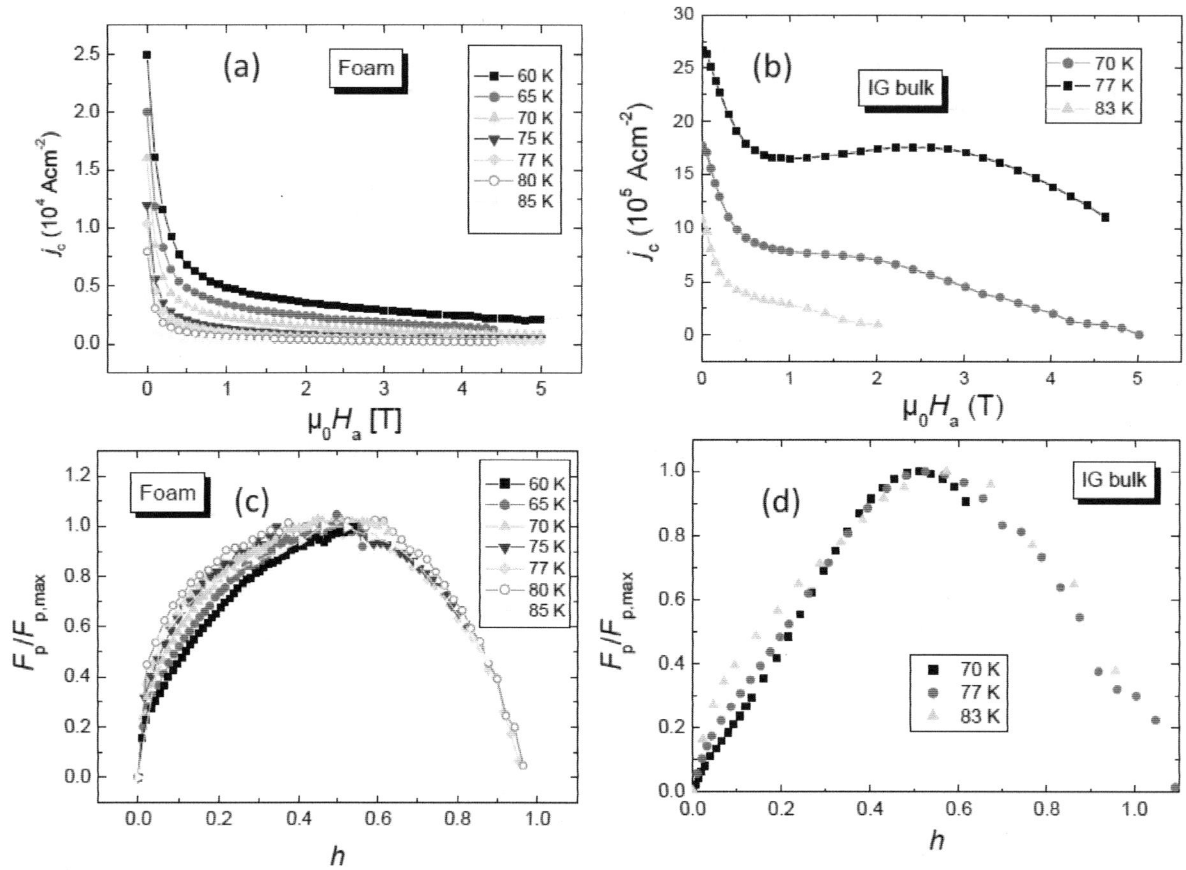

Figure 1. Critical current density for the foam (a) and the IG-processed sample (b) and the corresponding flux pinning force scalings (c,d). The currents of the foam sample are much smaller than that of the IG-processed sample, but the pinning force scaling is similar. Note the extremely high peak positions of the scaling for both samples at $h_0 = 0.5$ (foam strut) and 0.52 (IG-processed sample).

AFM scans were performed using Digital Instruments Nanoscope III and IV controllers in contact mode and tapping in ambient conditions. A Q-control unit was used to improve the signal-to-noise ratio in the tapping mode. The magnetic characterization was performed using SQUID magnetometers (QD MPMS 5 and 7XL). The EBSD analysis was carried out on a TSL OIM analysis unit [9] mounted in a JEOL SEM microscope (JSM 7000F).

3. Results and discussion

Flux pinning data (F_p) were calculated from the critical current densities, j_c, as function of applied field. The irreversibility field, H_{irr}, was determined for the IG-processed sample using a criterion of 10^4 A/cm^2; in case of the foam, H_{irr} was determined directly from the *M*(*H*) loops. A scaling of the F_p data was performed using the approach of Dew-Hughes [10,11]. The data for j_c and F_p are presented in Figs. 1 (a,c) for the foam strut and in Figs. 1 (b,d) for the IG-processed sample.

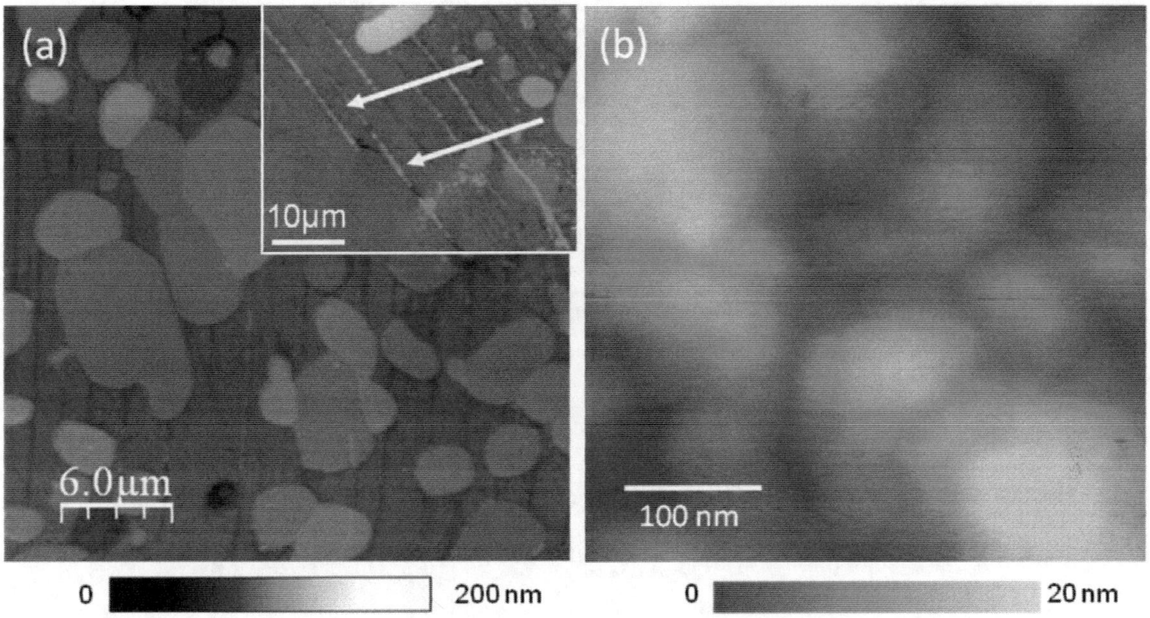

Figure 2. AFM topography images of polished surfaces from a foam strut (a) and the IG-processed YBCO sample (b). The inset to (a) shows the stripes filled with tiny secondary phase particles (arrows).

The $j_c(H)$-data of the foam show an exponential decay on increasing *H*, whereas $j_c(H)$ of the IG-processed sample reveals the peak effect. This is a direct consequence of the much easier oxygenation process of the open-cell foam structure. The $F_p(H)$-data of both samples are, however, pretty remarkable. The IG-processed sample reveals a good scaling, and the peak position, h_0, is located at a very high value of 0.52. Such a high value is reported in the literature only for the NEG-type 123-superconductors [12,13]. This is a consequence of the strong pinning of the δT_c-type which is also responsible for the peak effect. Another interesting result is the size of H_{irr} being extremely high for any type of YBCO sample measured so far. The magnetization data for the foam sample reveal homogeneous and symmetric *M*(*H*)-curves at low temperatures, but exhibit signs of granularity at elevated temperatures. Therefore, j_c reduces quickly at more elevated temperatures, and there is no peak effect. The critical currents determined for the foam sample are quite low as compared to the IG-sample, but similar values were found also previously in Refs. [14,15] at $T = 77$ K. Nevertheless, also for the foam sample the determined irreversibility fields are quite high.

Now, we have a closer look at the respective microstructures. Figure 2 presents AFM topography images of polished surfaces of both samples. The foam sample (a) exhibits many large, densely packed 211 particles embedded in the YBCO matrix. The grain size of the 211 particles ranges from 1 up to 10 μm. The YBCO matrix reveals characteristics stripes, being between 1 – 15 μm apart from each other. Topography images with higher magnification (see the inset and Ref. [7]) reveal that the spaces between these stripes are filled with tiny secondary phase particles with dimensions of about 200 nm. The EBSD imaging (Figs. 3,4) and EDX analysis demonstrates that these particles are 211 particles. In contrast to this, the 211 particles of the IG-processed sample are very small with the majority of them having sizes around 50 - 100 nm, but also some large 211 particles prevail.

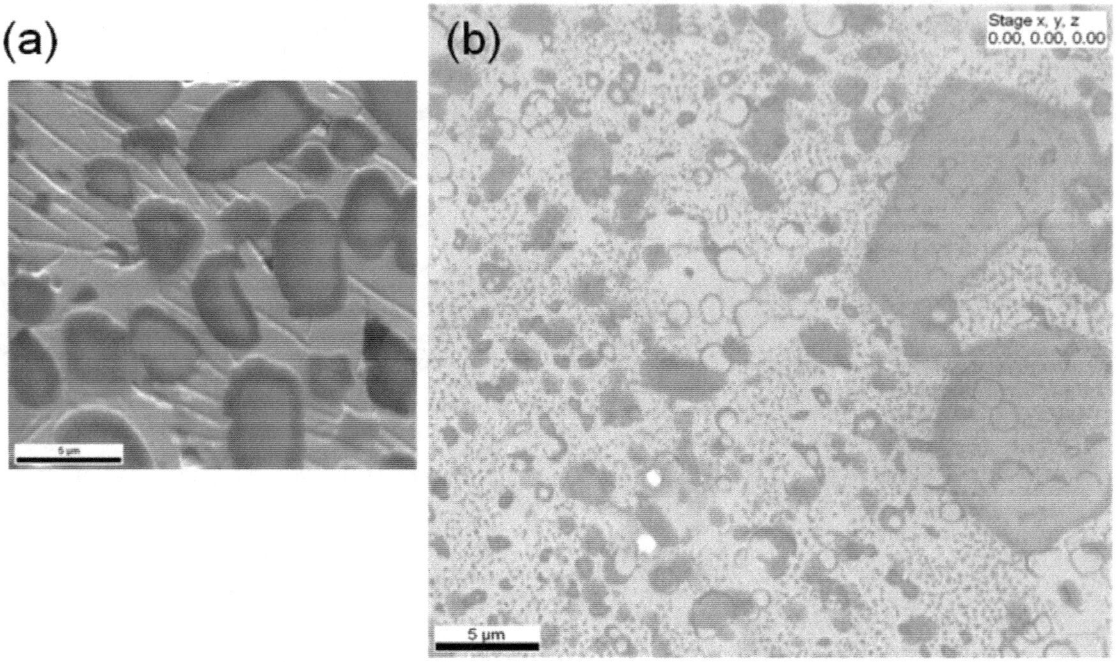

Figure 3. SEM images of the two sample types, taken in EBSD configuration (70° inclined to the electron beam). The two images are scaled so that the size of the scale bars is the same. (a) shows the foam strut, and (b) the IG-processed sample. The IG-processed sample exhibits a large number of tiny 211 particles embedded within the YBCO matrix. Note in both cases the smooth sample surface which is free of any scratches, indicating that the tiny 211 particles are formed naturally.

In order to learn more about the texture of the superconducting matrix and the influence of the embedded 211 particles on it, we performed an EBSD analysis on the samples. In this case, a two-phase scan is required. Figures 4 (a) and (b) show the phase maps of both samples; YBCO is denoted in red and the 211 particles in green. Additionally, the EBSD-detected grain boundaries are marked by black lines. In the foam strut large 211 particles dominate, but there are also many tiny 211 particles found within the YBCO matrix, and especially between the characteristic stripes of the YBCO phase as seen already by AFM. In the case of the IG-processed sample, the YBCO matrix is well extended (i.e., the entire section corresponds to one big grain) with 211 particles embedded within. However, there are also small YBCO subgrains formed close to the 211 particles and even within them. Figures 4 (c) and (d) present the IQ+IPF mappings (IQ = image quality parameter (gray scale) and IPF = inverse pole figure (color-coded, see the stereographic triangle)) in (001)-direction for the foam strut (a) and the IG-processed sample (b). In both cases, the 211 particles do not show a preferred orientation. The YBCO matrix of the foam strut (c) shows an orientation (the dominating color is orange, see the stereographic triangle), but also (110)-oriented areas (light blue) do exist. The phase map (a) of the foam strut reveals many small 211 particles being embedded within the stripes of the YBCO phase. However, the IPF map (c) shows that all these particles are randomly oriented. More details on the EBSD analysis of the foam sample are reported in Ref. [7]. The IG-processed sample shows a well-oriented YBCO matrix in (001)-orientation. Also in this case, the 211 particles are randomly oriented and exhibit internal misoriented subgrains.

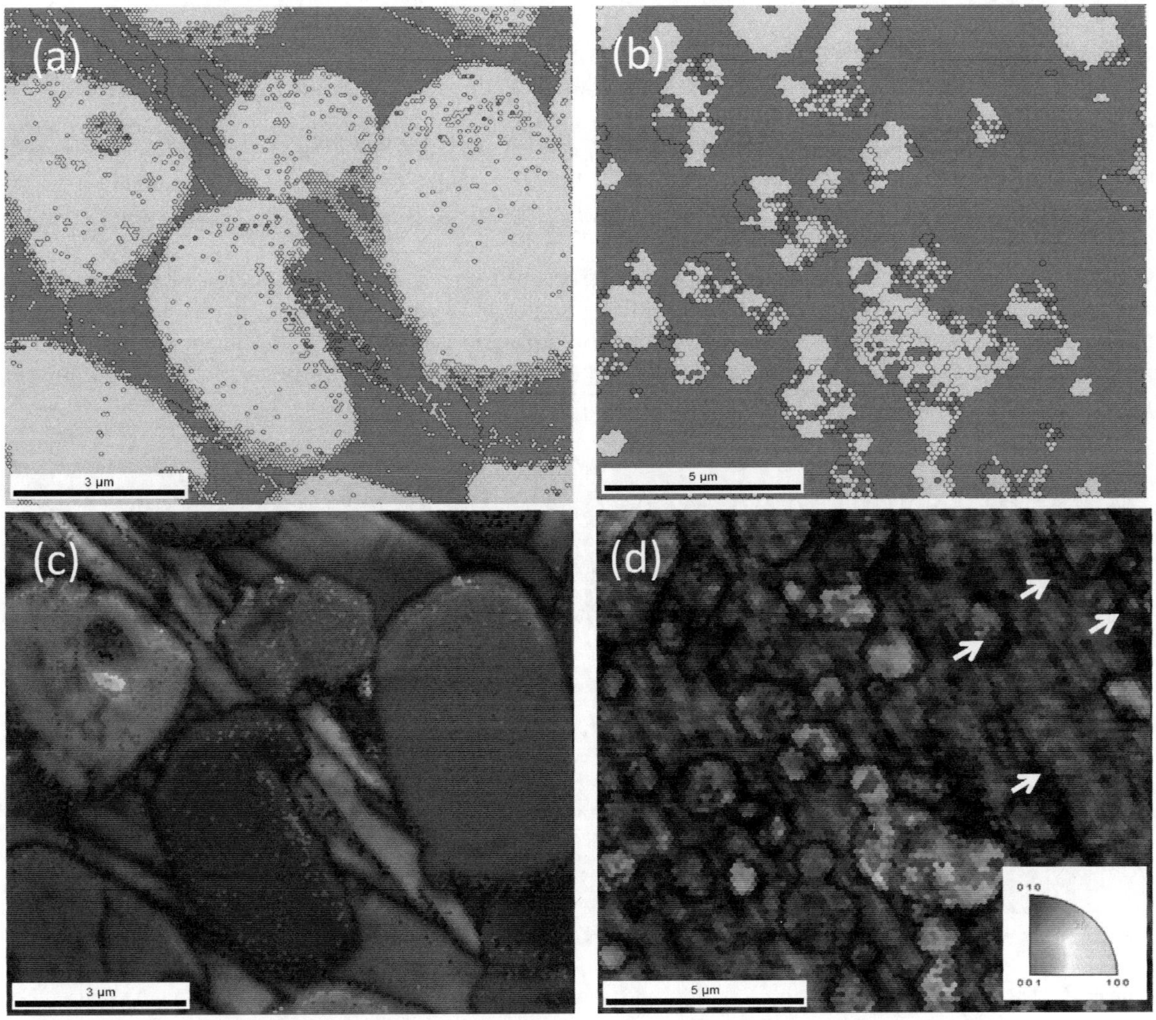

Figure 3. EBSD phase maps for the foam strut (a) and the IG-processed sample (b). YBCO is shown as red, and the 211 phase in green. In both maps, the detected grain boundaries are marked using black lines. EBSD IQ (gray scale)+IPF (color coded) maps in (001)-orientation for the foam strut are shown in (c) and the same for the IG-processed sample in (d). The arrows in (d) mark tiny 211 particles embedded in the YBCO matrix. The color code for the IPF maps is given in the stereographic triangle, which is the same for both phases.

A comparison of the phase map and the IPF map reveals that in the close neighborhood of the larger 211 particles also misoriented YBCO particles exist. This is due to a mismatch of the unit cells of the two phases as already observed in Ref. [16]. At some orientations of the 211 phase, the growth of the YBCO phase is altered due to stress/strain, and subgrains are formed. The tiny 211 particles are not detected in this EBSD scan due to the stepsize of 70 nm used for the EBSD mapping. Only some ~100 nm particles were found (marked by arrows) within the YBCO matrix. Therefore, an EBSD scan with a smaller stepsize is required to detect the tiny 211 particles embedded in the matrix. The high image quality of the Kikuchi patterns achieved in the previous scans allows such measurement.

Figures 5 (a,b) present an EBSD measurement on the IG-processed sample with a stepsize of 40 nm. The large number of grain boundaries seen in (a) reveals that the YBCO phase exhibits misoriented grains with sizes ranging between 80 and 800 nm. The grain size of the majority of the 211 particles is between 50 and 100 nm, but also some larger 211 particles (lower right corner) can be observed. The IQ+IPF map (b) shows some interesting features: (i) There are only two orientations of the 211 particles (green, blue) found in the entire scan area and also of the YBCO subgrains, which demonstrates the alignment of the YBCO and 211 particles due to stress/strain as discussed before. (ii) The comparison of the phase map and the IQ+IPF-map reveals that on this scale, there is a large amount of misorientation of the YBCO phase due to the presence of the 211 particles. These misorientations of the YBCO subgrains together with the tiny 211 particles provide an unique flux pinning landscape, which is directly responsible for the large irreversibility fields observed in the magnetic measurements. (iii) The image quality of such a high resolution scan is much lower as compared to Fig. 4. This shows that these measurements are close to the detection limit of the standard EBSD technique on samples with nano-sized structures; a higher resolution can only be achieved employing the transmission EBSD (t-EBSD) technique [17].

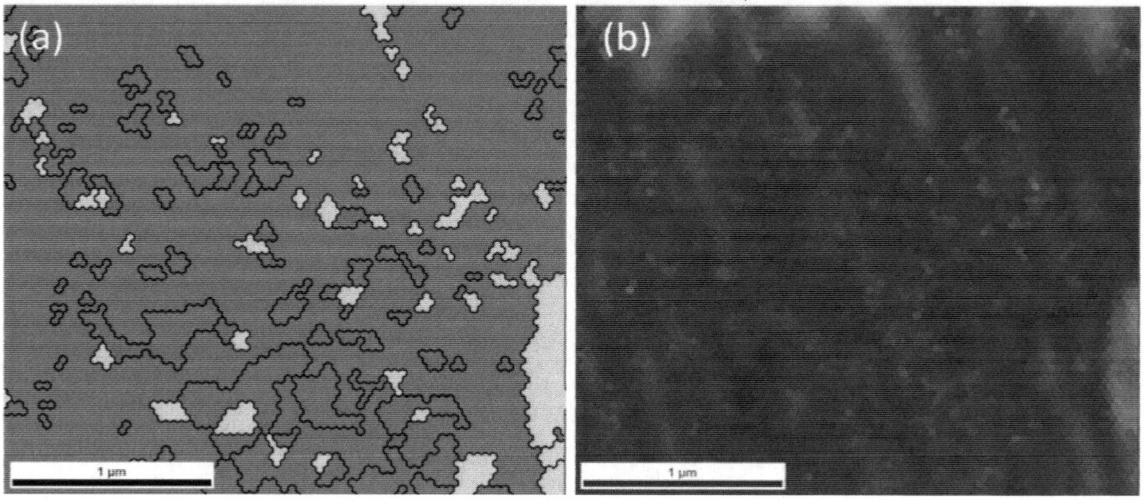

Figure 5. EBSD measurements on the IG-processed sample with a stepsize of 40 nm. (a) gives the phase map (YBCO – red, 211 – green) and the detected grain boundaries are marked with black lines. (b) shows the IQ+IPF map in (001)-direction. The color code is the same as in Fig. 3 (d).

To summarize our observations on both types of samples, the microstructure analysis clearly reveals the differences, but also the common features of the two types of materials produced by the IG-technique. While the overall orientation of the YBCO matrix is different in the two samples, the presence of tiny 211 particles is common for both and according to the results presented in Ref. [13], directly responsible for the high irreversibility fields due to the effective flux pinning. The high irreversibility fields and the high peak positions in the pinning force scaling make the IG-processed materials ideal candidates for high-field applications.

Conclusions

The comparison of a YBCO foam sample with an IG-processed YBCO bulk reveals many differences, but also common features due to the similar processing technique. The foam contains many large 211 particles, but also tiny ones which provide additional flux pinning sites. The YBCO matrix is oriented, but not as perfect as in the case of the IG-processed bulk. Here, the 211 particle size is much smaller, and the matrix contains even smaller (~50 nm) 211 particles. EBSD scans with high resolution (stepsize 40 nm) reveal that the 211 particles only show two main orientations, which cause misorientations of the YBCO matrix. This characteristic pinning landscape is essential for the high irreversibility fields observed.

Acknowledgements

We would like to thank G J Schmitz (ACCESS, RWTH Aachen) for the foam sample and J. Schmauch (Saarland University) for his technical assistance.

References

[1] Reddy E S, Herweg M and Schmitz G J 2003 *Supercond. Sci. Technol.* **16** 608

[2] Reddy E S and Schmitz G J *Am. Ceram. Soc Bull.* **81** 35

[3] Iida K, Hari Babu N, Shi Y and Cardwell D A 2005 *Supercond. Sci. Technol.* **18** 1421

[4] Li G Z, Li D J, Deng X Y, Deng J H and Yang W M 2013 *Cryst. Growth Des.* **13** 1246

[5] Nakazato K, Muralidhar M. Koblischka M R and Murakami M 2014 *Cryogenics* **63** 129

[6] Nakazato K, Muralidhar M, Koshizuka N, Inoue K and Murakami M 2014 *Physica C* **504** 4
[7] Koblischka M R, Koblischka-Veneva A, Reddy E S and Schmitz G J 2014 *J. Adv. Ceram.* **3** 317
[8] Koblischka M R and Koblischka-Veneva A 2013 *Phase Transitions* **86** 651
[9] TexSEM Laboratories (TSL), Draper, UT, USA, 2004
[10] Dew-Hughes D 1974 *Phil. Mag.* **30** 293
[11] Koblischka M R, van Dalen AJJ, Higuchi T, Yoo S I and Murakami M 1998 *Phys. Rev.* B **58** 2863
[12] Muralidhar M, Koblischka M R and Murakami M 2000 *Inst. Phys. Conf. Ser.* **167** 155
[13] Koblischka M R and Murakami M 2000 *Supercond. Sci. Technol.* **13** 738
[14] Noudem J G, Reddy E S and Schmitz G J 2003 *Physica C* **390** 286
[15] Bartolomé E, Granados X, Puig T, Obradors X, Reddy E S and Schmitz G J 2004 *Phys. Rev. B* **70** 144514
[16] Koblischka-Veneva A, Koblischka M R, Mücklich F, Ogasawara K and Murakami M 2005 *Supercond. Sci. Technol.* **18** S158
[17] Trimby P W 2012 *Ultramicroscopy* **120** 16

Influence of crossed fields in structures combining large grain, bulk (RE)BCO superconductors and soft ferromagnetic discs

M P Philippe[1], J F Fagnard[1], L Wéra[1], M Morita[2], S Nariki[2], H Teshima[2], H Caps[3], B Vanderheyden[1] and P Vanderbemden[1]

[1] University of Liège, Department of Electrical Engineering & Computer science (B28), Sart-Tilman, B-4000 Liège, Belgium
[2] Nippon Steel & Sumitomo Metal Corporation, Shintomi 20-1, Futtsu, Chiba 293-8511, Japan
[3] University of Liège, GRASP, Physics Department (B5), B-4000 Liège, Belgium

E-mail : Philippe.Vanderbemden@ulg.ac.be

Abstract. Bulk (RE)BCO superconductors are able to trap record magnetic fields and can be used as powerful permanent magnets in various engineering applications such as rotating machines and magnetic bearings. When such superconducting (SC) "trapped field magnets" are combined to a ferromagnetic (FM) disc, the total magnetic moment is increased with respect to that of the superconductor alone. In the present work, we study experimentally the magnetic behaviour of such hybrid FM/SC structures when they are subjected to cycles of applied field that are orthogonal to their permanent magnetization, i.e. a "crossed-field" configuration. Experimental results show that the usual "crossed-field demagnetization" caused by the cycles of transverse field is strongly reduced in the presence of the ferromagnet.

1. Introduction

Due to their strong flux pinning ability, bulk (RE)$Ba_2Cu_3O_{7-\delta}$ (REBCO) superconductors (where "RE" denotes a rare-earth ion) form the basis of "trapped field superconducting magnets" with unprecedented characteristics [1-5]. They can be used in a range of power applications including levitation guideways or rotating machines [6-8]. Several physical phenomena, however, cause the trapped flux density generated by such superconducting permanent magnets to decrease after they have been magnetized. The first is flux creep, which usually leads to a logarithmic time-dependence decay of the magnetization [9,10]. The second is related to the interaction of the bulk superconductor with the magnetic fields required for the operation of the device. As an example, the superconductor placed in the rotor of a synchronous motor is subjected to the rotating field generated by the stator [11]. Due to transients that can arise e.g. during a sudden change of mechanical load in the device, the superconductor is likely to be subjected to parasitic time-varying fields [12,13]. In addition to causing losses and self-heating due to vortex motion [14,15], these parasitic fields may substantially reduce the trapped flux density in the superconductor when their direction is orthogonal to that of the main magnetization [16,17]. This phenomenon, called "crossed-field demagnetization" or "collapse of the magnetic moment" has been studied for a long time in superconducting materials of various types [17-22] and is still an intense subject of research [23-25] due to the complexity of physical phenomena involved [26,27].

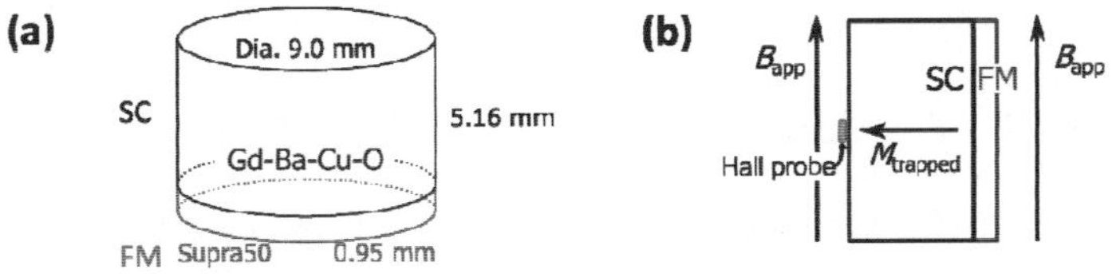

Figure 1. (a) Schematic representation of a superconductor/ferromagnet (SC/FM) hybrid structure made of a 5.16 mm thick $GdBa_2Cu_3O_7$ superconductor attached to a 0.95 mm thick ferromagnetic disc made of "*Supra50*" material (b) Schematic representation of the hybrid SC/FM structure subjected to a crossed-field experiment. $M_{trapped}$ is the permanent magnetization trapped by the superconductor. B_{app} represents the transverse magnetic flux density, applied perpendicularly to $M_{trapped}$. The Hall probe is used to measure the axial component of the flux density, i.e. along the direction of $M_{trapped}$.

Recently, it was shown both experimentally and numerically that the magnetic performances of trapped field superconducting (SC) magnets can be enhanced by combining them to a short section of soft ferromagnetic (FM) material, thereby realizing a hybrid superconductor / ferromagnet (SC/FM) structure [28,29]. Up to now, the magnetic properties of these structures were investigated for a range of applied fields applied *parallel* to the trapped magnetization. The aim of the present work is to study experimentally the properties of such SC/FM structures in the crossed configuration, i.e. when the applied field is *perpendicular* to the trapped magnetization. The corresponding decay of flux density at the surface of the hybrid SC/FM structure will be compared to that occurring at the surface of a superconductor alone. This will allow us to draw conclusions about the practical use of such hybrid SC/FM structures used as trapped field magnets in engineering applications.

2. Experiment

Crossed-field measurements are carried out on bulk, large grain $GdBa_2Cu_3O_7$ (Gd-Ba-Cu-O) superconductor from Nippon Steel & Sumitomo Metal Corporation. The studied superconducting sample is a disk of 9 mm in diameter and 5.16 mm in thickness. The hybrid SC/FM structure is made by attaching a 0.95 mm thick ferromagnetic disk of the same diameter as the superconductor, as shown in figure 1(a). The soft ferromagnetic material is "*Supra50*" from Aperam. Its magnetic properties were characterized independently in an earlier work [28].

The magnetic properties of the $GdBa_2Cu_3O_7$ superconductor (SC) and the hybrid (SC/FM) structure are first characterized at 77 K in a bespoke magnetometer designed to accommodate large bulk samples up to 17 mm in diameter. The design, physical operation, and calibration of the experimental system are described extensively in a previous work [30]. The magnetometer is able to measure magnetic moments in excess of 1 Am^2 (1000 emu), which is two orders of magnitude above the maximum magnetic moment of commercial cryogenic measurement systems. The samples are first magnetized by a field-cooled (FC) process down to $T = 77$ K under 0.625 T in a separate electromagnet. Then the magnetic moment is recorded over time using the magnetometer.

The crossed-field behaviour is then studied at 77 K using a custom probe with a rotating sample holder that can be inserted in the experimental chamber of a Quantum Design Physical Property Measurement System (PPMS). The system includes a AREPOC HPP-VU Hall probe placed against the superconductor. The Hall probe is used to record the axial flux density at the centre of the circular face, as shown in figure 1(b). A field-cooled (FC) procedure under a constant 1 T background axial field is used to magnetize permanently the superconductor or the hybrid structure. After the magnetizing field is switched off, the sample is rotated by 90°; the axial trapped flux density at the surface of the sample is measured by the Hall probe for 1500 s to record the natural relaxation of the sample magnetization caused by flux creep effects. Then 10 cycles of magnetic field perpendicular to the trapped flux density are applied following a triangular waveform at a sweep rate of 3.3 mT/s. Two sets of experiments with different amplitudes (50 mT and 100 mT) are carried out. It should be noted that the mechanical components of the rotating sample holder are designed carefully to withstand the possibly high magnetic torque resulting from the interaction between the magnetized sample and the externally applied transverse magnetic field [31].

3. Results and discussion

Figure 2 compares time-dependence of the trapped magnetic moment of the bulk superconducting sample (SC) and that of the hybrid (SC/FM) structure at $T = 77$ K. The elapsed time t corresponds to that after the magnetization process. The red dashed lines in figure 2 are a linear regression of the data in a log-log scale. As can be seen, the magnetic moment of both samples exhibits a power law behaviour. Fitting the experimental magnetic moment to the theoretical law [9]

$$m(t) = m_0 \left(1 + \frac{t}{t_0}\right)^{\frac{1}{(1-n)}}$$

allows one to obtain the initial magnetic moment m_0 and the n value of the $E \sim J^n$ power law. For the $GdBa_2Cu_3O_7$ superconductor (SC), this procedure leads to $m_0 = 0.12$ Am² and $n = 32.9$. If a field-independent critical current density is assumed and, at first approximation, finite size effects are neglected, J_c can be estimated from the magnetic moment m_0 using $J_c \approx 3m_0/(aV)$, where a is the sample radius and V its volume. This gives $J_c \approx 2.44 \times 10^8$ A/m², which is amongst the highest reported values for large grain, bulk (RE)BCO superconductors [4]. The same procedure applied on the (SC/FM) hybrid structure gives $m_0 = 0.13$ Am² and $n = 32.1$. The 8 % increase of magnetic moment is in excellent agreement with previous observations and calculations made on other hybrid structures involving $YBa_2Cu_3O_7$ superconductors with ferromagnetic discs of different thicknesses [28,29]. In addition, the relaxation behaviour of the trapped magnetic moment of the hybrid structure is almost unaffected by the presence of the soft ferromagnet.

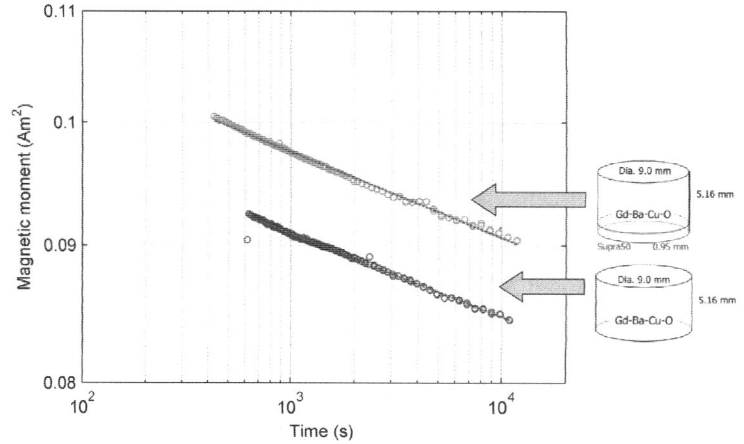

Figure 2. Magnetic moment vs. time for the bulk superconductor alone (blue) or for the hybrid superconducting/ferromagnet structure (green). The red dashed lines are obtained by linear regression on the data in a double logarithmic scale.

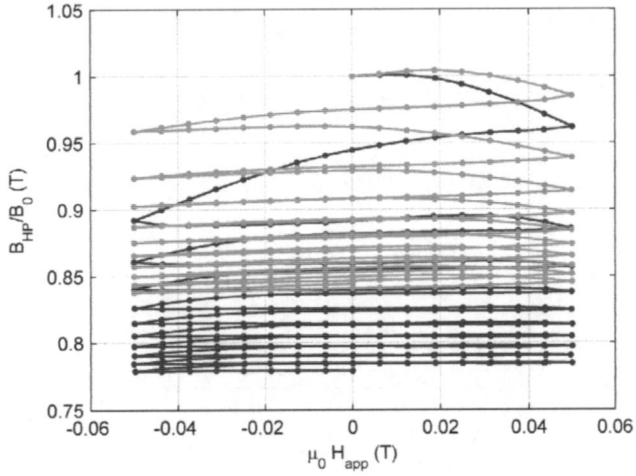

Figure 3. Decrease of the axial component of the flux density (normalized to the flux density before applying the transverse field) measured by the Hall probe B_{HP} during application of 10 cycles of transverse fields, as shown in figure 1(b). The measurement is carried out on the superconductor alone (blue) and on the SC/FM hybrid structure (green). The measurement temperature is 77 K.

Now we examine the results obtained in the crossed-field configuration. Figure 3 shows the flux density measured by the Hall probe against the sample surface during 10 cycles of the crossed field (amplitude = 50 mT) for the superconductor alone (blue) and for the hybrid structure (green). For clarity, data are normalized with respect to the flux density before applying the transverse field (B_0). For both samples, the largest decrease occurs during the first cycle. Remarkably, the hybrid structure is much less affected by the transverse field than the superconductor alone. The decrease after the first cycle of transverse field is ~ 4% for the hybrid structure, against ~ 11% for the superconductor alone. This behaviour is confirmed for the next cycles: after 10 cycles, the total decrease is ~ 22% for the superconductor only, while it is only ~ 16% for the hybrid structure. Figure 4 compares the normalized flux density as a function of crossed-field cycles for two transverse field amplitudes of 50 mT and 100 mT. In both cases, the demagnetization is seen to be appreciably smaller for the case of the hybrid structure. Note also that the relatively large decays observed in figures 3 and 4 are measured at 77 K. In engineering applications where the operating temperature is in the range of 30 - 50 K, the critical current and the creep exponent are higher than at 77 K and the overall decay of the magnetization for a given crossed field amplitude is much smaller than those observed in the present work.

The above results point out the beneficial effect of a ferromagnetic piece on one side of the superconductor to reduce the crossed-field attenuation of the available flux density. These results can be understood by the shielding effect provided by the ferromagnet. Due to the low reluctance path, a part of the transverse flux density lines are expected to be diverted from the bulk superconductor, thereby minimizing the redistribution of persistent currents that usually takes place in the presence of crossed fields [20-22]. In the present case, it should also be emphasized that the self-heating of the superconductor caused by the time-varying transverse field can be neglected because of the low-value of the sweep rate used [14,25].

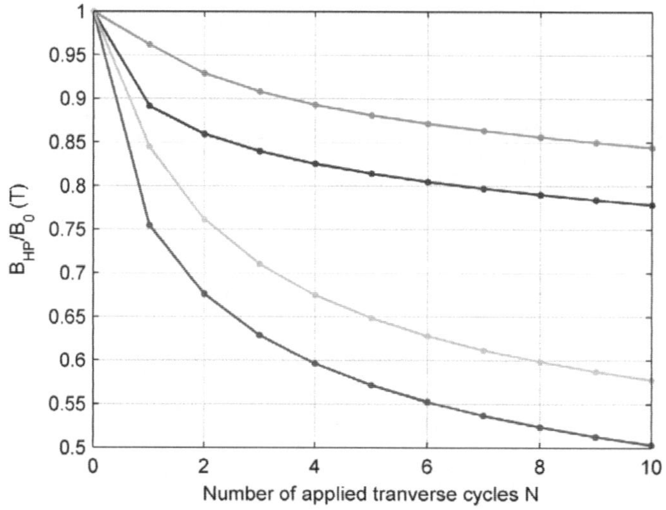

Figure 4. Magnetic flux density B_{HP} measured by the Hall probe as a function of the number of full transverse applied field cycles. For a cycle amplitude of 50 mT, the measurements are in blue and green for the SC alone and SC/FM structure, respectively. The corresponding data recorded at 100 mT for the SC alone and SC/FM structure are in red and yellow, respectively.

5. Conclusion

In this work we investigated the effect of a ferromagnetic disc placed against a bulk $GdBa_2Cu_3O_7$ superconductor when it is first fully magnetized and then subjected to various cycles of transverse field perpendicular to the trapped magnetic moment. The results give evidence that the addition of a ferromagnetic disc on one side of the superconductor reduces the collapse of the trapped flux density against the other face of the superconductor. Therefore one can conclude that the presence of the ferromagnet is not only helpful in increasing the trapped magnetic moment but also to reduce the detrimental effect of crossed fields. Such a result is important for use of bulk melt-textured superconductors in engineering applications.

Acknowledgements

The research was funded through the University of Liège (ULg) and an ARC grant for Concerted Research Actions, financed by the French Community of Belgium (Wallonia-Brussels Federation), under reference ARC 11/16-03.

References

[1] Durrell J H *et al.* 2014 *Supercond. Sci. Technol.* **27** 082001
[2] Tomita M and Murakami M 2003 *Nature* **421** 517
[3] Teshima H, Morita M and Hirano H 2006 *J. Phys.: Confer. Series* **43** 1031
[4] Devendra Kumar N, Rajasekharan T, Muraleedharan K, Banerjee A and Seshubai V 2010 *Supercond. Sci. Technol.* **23** 105020
[5] Doyle R A, Bradley A D, Lo W, Cardwell D A, Campbell A M, Vanderbemden P and Cloots R 1998 *Appl. Phys. Lett.* **73** 117
[6] Werfel F N, Floegel-Delor U, Rothfeld R, Riedel T, Goebel B, Wippich D and Schirrmeister P 2012 *Supercond. Sci. Technol.* **25** 014007

[7] Zhen Huang, Wei Xian, Min Zhang, Chudy M, Chen Y, Zhaoyang Zhong, Baghdadi M, Wei Wang, Spaven F, Matsuda K and Coombs T A 2013 *IEEE Trans. Appl. Supercond.* **23** 5200204
[8] Zhou D, Izumi M, Miki M, Felder B, Ida T and Kitano M 2014 *Supercond. Sci. Technol.* **25** 103001
[9] Gurevich A and Küpfer H 1993 *Phys. Rev. B* **48** 6477
[10] Cardwell D A *et al.* 2005 *Supercond. Sci. Technol.* **18** S173
[11] Campbell A M 2014 *Supercond. Sci. Technol.* **27** 124012
[12] Qiu M, Huo H K, Xu Z, Xia D, Lin L Z and Zhang G M 2005 *IEEE Trans. Appl. Supercond.* **15** 3172
[13] Yamagishi K, Ogawa J and Tsukamoto O 2015 *IEEE Trans. Appl. Supercond.* **25** 5201905
[14] Vanderbemden P, Laurent P, Fagnard J F, Ausloos M, Hari Babu N and Cardwell D A 2010 *Supercond. Sci. Technol.* **23** 075006
[15] Zushi Y, Asaba I, Ogawa J, Yamagishi K, Tsukamoto O, Murakami M and Tomita M 2004 *Physica C* **412-414** 708
[16] Ogawa J, Iwamoto M, Yamagishi K, Tsukamoto O, Murakami M and Tomita M 2003 *Physica C* 386 26
[17] Vanderbemden P, Hong Z, Coombs T A, Ausloos M, Hari Babu N, Cardwell D A and Campbell A M 2007 *Supercond. Sci. Technol.* **20** S174
[18] Funaki K and Yamafuji K 1982 *Jpn. J. Appl. Phys.* **21** 299
[19] Fisher L M, Kalinov A V, Savel'ev S E, Voloshin I F, Yampol'skii V A, LeBlanc M A R and Hirscher S 1997 *Physica C* **278** 169
[20] Badía A, and López C 2007 *Phys. Rev. B* **76** 054504
[21] Vanderbemden P, Hong Z, Coombs T A, Denis S, Ausloos M, Schwartz J, Rutel I B, Hari Babu N, Cardwell D A and Campbell A M 2007 *Phys. Rev. B* **75** 174515
[22] Luzuriaga J, Badia-Majos A, Nieva G, Giordano J L, Lopez C, Serquis A and Serrano G 2009 *Supercond. Sci. Technol.* **22** 015021
[23] Baghdadi M, Ruiz H S and Coombs T A 2014 *Appl. Phys. Lett.* **104** 232602
[24] Celebi S, Sirois F and Lacroix C 2015 *Supercond. Sci. Technol.* **28** 025012
[25] Fagnard J F, Kirsch S, Morita M, Teshima H, Vanderheyden B and Vanderbemden P 2015 *Physica C* **512** 42
[26] Campbell A M 2011 *Supercond. Sci. Technol.* **24** 091001
[27] Mikitik G P and Brandt E H 2004 *Phys. Rev. B.* **69** 134521
[28] Philippe M P, Fagnard J F, Kirsch S, Xu Z, Dennis A R, Shi Y H, Cardwell D A, Vanderheyden B and Vanderbemden P 2014 *Physica C* **502** 20
[29] Philippe M P, Ainslie M D, Wéra L, Fagnard J F, Dennis A R, Shi Y H, Cardwell D A, Vanderheyden B and Vanderbemden P 2015 *Supercond. Sci. Technol.* **28** 095008
[30] Egan R, Philippe M, Wera L, Fagnard J F, Vanderheyden B, Dennis A, Shi Y, Cardwell D A and Vanderbemden P 2015 *Rev. Sci. Instrum.* **86** 025107
[31] Koblischka M R, van Dalen A J J and Ravi Kumar G 1996 *J. Supercond.* **9** 143

Electric transport measurements on bulk, polycrystalline MgB_2 samples prepared at various reaction temperatures

A Wiederhold[1], M R Koblischka[1], K Inoue[2], M Muralidhar[2], M Murakami[2], and U Hartmann[1]

[1]Institute of Experimental Physics, Saarland University, Campus C 6 3, D-66123 Saarbrücken, Germany.
[2]Superconducting Materials Laboratory, Department of Materials Science and Engineering, Shibaura Institute of Technology 3-7-5 Toyosu, Koto-ku, Tokyo 135-8548, Japan.

a.Wiederhold@physik.uni-saarland.de

Abstract. A series of disk-shaped, bulk MgB_2 superconductors (sample diameter up to 4 cm) was prepared in order to improve the performance for superconducting super-magnets. Several samples were fabricated using a solid state reaction in pure Ar atmosphere from 750 to 950 °C in order to determine the optimum processing parameters to obtain the highest critical current density as well as large trapped field values. Additional samples were prepared with added silver (up to 10 wt.-%) to the Mg and B powder. Magneto-resistance data and *I/V*-characteristics were recorded using an Oxford Instruments Teslatron system. From Arrhenius plots, we determine the TAFF pinning potential, U_0. The *I/V*-characteristics yield detailed information on the current flow through the polycrystalline samples. The current flow is influenced by the presence of pores in the samples. Our analysis of the achieved critical currents together with a thorough microstructure investigation reveals that the samples prepared at temperatures between 775 °C and 805 °C exhibit the smallest grains and the best connectivity between them, while the samples fabricated at higher reaction temperatures show a reduced connectivity and lower pinning potential. Doping the samples with silver leads to a considerable increase of the pinning potential and hence, the critical current densities.

1. Introduction

The metallic superconductor MgB_2, which was discovered 2001, has a critical temperature T_c of 39 K [1]. Even though this is low, compared to the T_c of high temperature superconductors, it posses several advantages. Due to its small coherence length it does not suffer from the weak link problem, wherefore it is possible to create cheap, large sized polycrystalline bulk samples [2]. Besides it has a high current transport capability [3] and strong trapped fields [4], which makes it to an ideal material for practical applications [5] like trapped field magnets. Therefore it is important to work out a cost-effective preparation method like simple sintering, which enables to produce large bulk samples with minimal cost [6-9]. In order to optimize the processing conditions, a series of samples was fabricated at various reaction temperatures ranging from 775 °C to 950 °C as described in Refs. [10,11]. For this series of samples, the magnetic and electric properties were studied in detail to evaluate the critical current densities, the trapped fields, the effective flux pinning mechanism and the grain connectivity. Due to the porous structure of the sintered material, an additional set samples with an addition of silver (up to 10 wt.-%), which was already successfully used to improve the mechanical abilities and j_c values of (RE)BCO materials [13,14], were produced to improve the mechanical properties and the same measurements were performed there. The magnetic data were already presented in Ref. [12], so we focus in this contribution on the electric properties (resistance and *I*/*V*-characteristics) of this set of samples.

1. Experimental procedure

Polycrystalline MgB_2 samples were fabricated by using in-situ solid state reaction. Mg and B powders of high purity were pressed into pellets and sintered under argon atmosphere at temperatures from 775 to 950 °C. The silver doped samples were produced in the same way by adding up to 10 wt.-% of silver powder. Here the sintering temperature was selected to be 775 °C. More details on the sample preparation and their characterization are given elsewhere [10-12].
For the transport measurements, long bars with dimensions $1 \times 2 \times 8$ mm^3, resulting in a distance of 5 mm between the voltage contacts, were cut from bulk MgB_2 samples using a diamond saw. The resistance was measured in four-probe configuration with a dc current of 30 mA applied perpendicular to the magnetic field direction. Thin copper wires of 100 µm in diameter were used as leads for low resistance contacts placed onto the samples by means of silver paste.

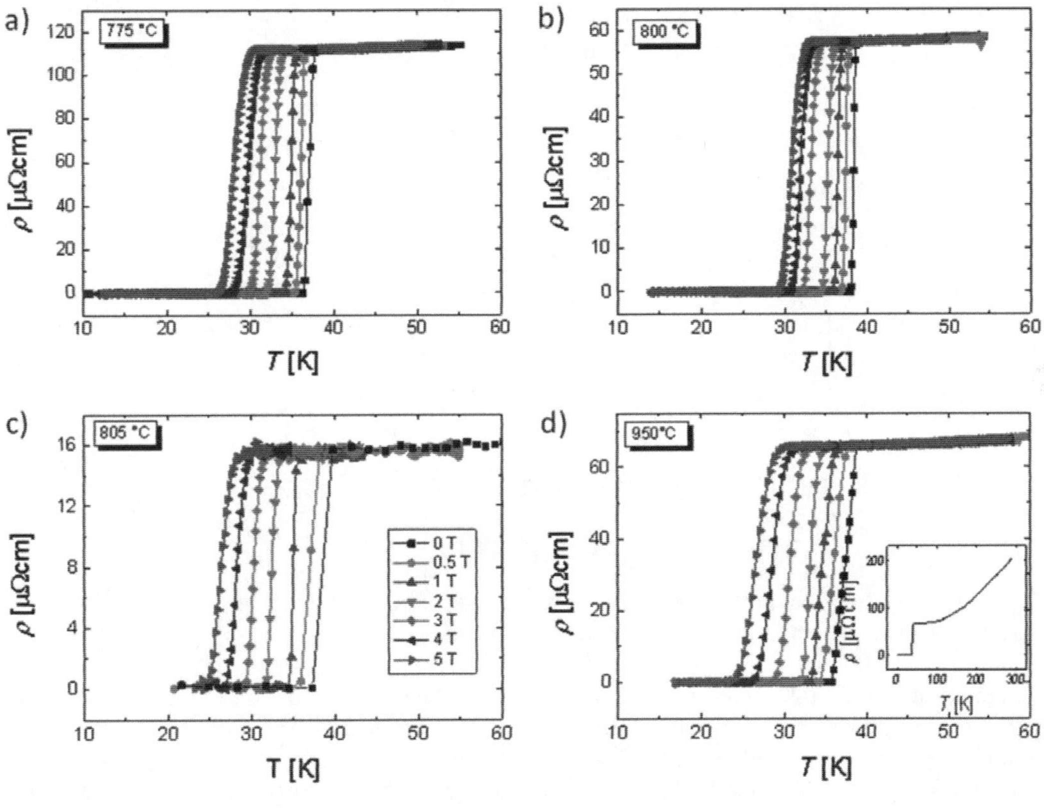

Figure 1. Electrical resistivity as function of temperature for different reaction temperatures measured in magnetic fields up to 5 T. The curves were obtained at a current of 30 mA. The inset in (d) presents the full curve from 4 to 300 K in zero field for the sample sintered at 950 °C.

Field steps of 1 T were used for the *R*(*T*) measurements, and field steps of 0.25 T to collect the *E*-*j*-data. The magnetic field sweep rate was 0.7 T/min, and temperature steps of 0.25 K were applied.
The electric measurements were performed in an Oxford Instruments 10/12 T Teslatron cryostat in the temperature range 4 K – 300 K. The measurements were controlled by a MatLab program operating a stabilized current source (Keithley 2400) and a voltmeter (Keithley 2100 multimeter) [15].

2. Results and discussion

Figures 1 (a)-(d) present the resistivity data of the MgB_2 samples prepared at 775 °C, 800 °C, 805 °C and 950 °C. These samples were selected on the basis of the magnetic measurements performed in Refs. [5,7]. The first three samples were showing the best magnetic properties (critical current density, j_c, irreversibility fields and trapped fields), and the sample prepared at 950 °C showed strong pinning due to embedded MgB_4 particles [10,12].
All samples show a sharp superconducting transition which is shifting to lower values on increasing the external magnetic fields. With the electric resistivities in the normal conducting state, the geometrical connectivity can be estimated with $K = \Delta\rho_g/\Delta\rho$, while $\Delta\rho_g$ can be assumed to be 6.31 μΩcm for randomly orientated three dimensional samples and $\Delta\rho = \rho(300\ K) - \rho(40\ K)$ [16]. The measured values for the normal conducting resistivities are shown in table 1. Figure 2 shows the calculated values for the connectivity as a function of the the sintering temperature. The connectivity shows an increase with growing processing temperature until a maximum at 805 °C, and then it decreases again, which reveals well coupled grains for sintering temperatures around 805 °C.

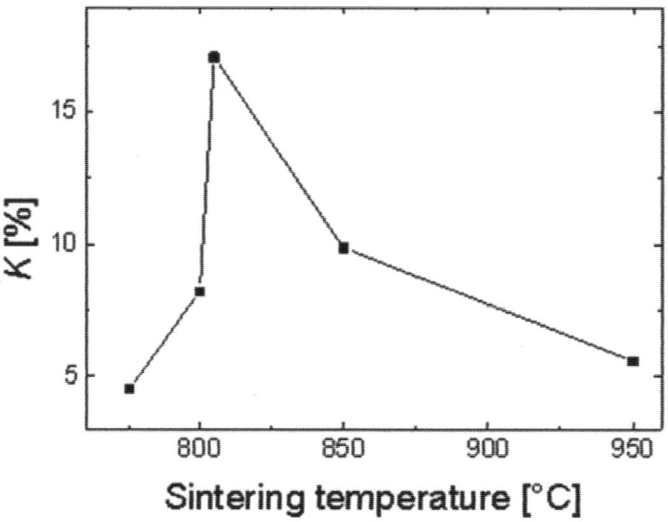

Figure 2. Connectivity as function of sintering temperature. The maximum at 805 °C indicates well coupled grains for process temperatures around this point.

The data can be converted into Arrhenius-plots by plotting the logarithm of the electric resistivity against the inverse temperature, which is shown in figures 3 (a)-(d). From the slopes of the linear parts, the TAFF pinning potential can be calculated [17]. The *V*-*I* curves of the samples sintered at 775 °C, 800 °C, 805 °C and 950 °C are presented in figures 4 (a)-(d). As in the case discussed above, the critical current density shifts to lower values with increasing magnetic fields. Also a change in the shape of the curves can be observed on increasing reaction temperature. These curves enable the pinning potential in the flux creep region to be calculated [18].

Sintering temperature [°C]	ρ(300 K) [μΩcm]	ρ(40 K) [μΩcm]	$\Delta\rho$ [μΩcm]
775	251	111	140
800	134	57	77
805	52	15	37
850	106	42	64
950	179	66	113

Table 1. Normal conducting resistivity at 300 and 40 K and the calculated phonon term resistivity $\Delta\rho$ for the samples sintered at different temperatures.

Figure 5 shows the TAFF pinning potential calculated from the resistance data (a) and the flux creep pinning potential calculated from the *V-I* curves. The TAFF pinning potentials show strong pinning centers for the samples sintered at around 805 °C which is in agreement with Ref. [19]. In this study magnetic measurements were performed on the same samples and the highest j_c value of about 250 kA/cm² was observed at a sintering temperature of 805 °C. The flux creep potentials show similar results.

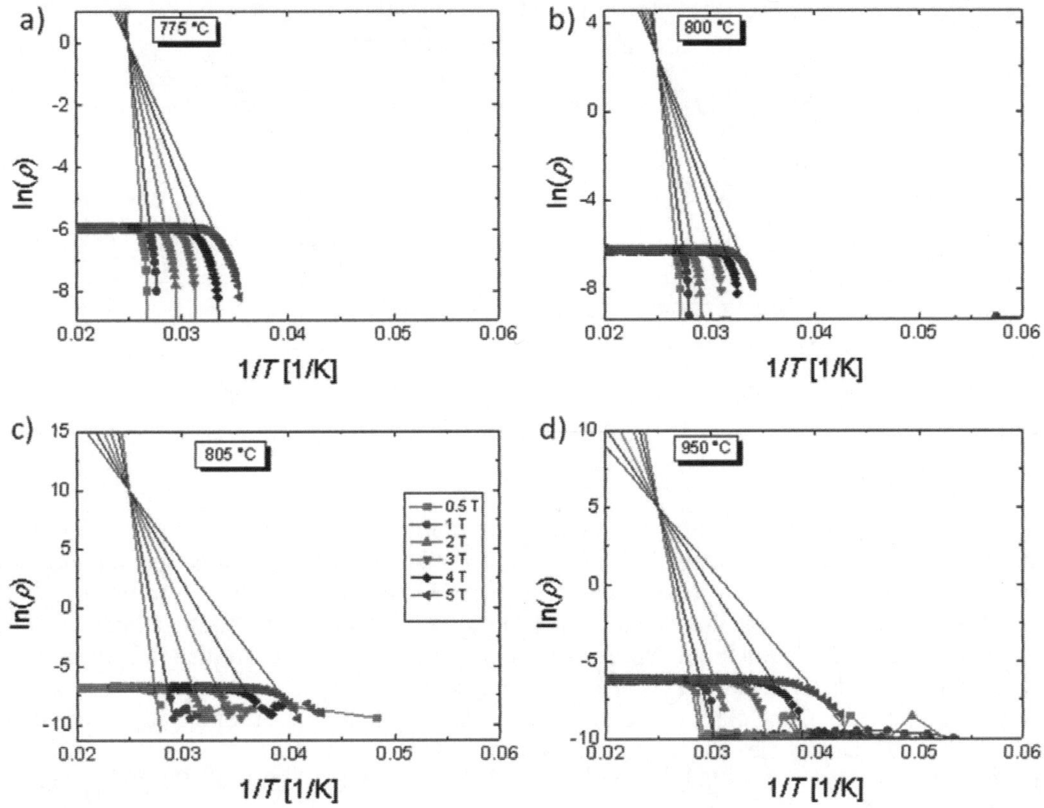

Figure 3. Arrhenius-plots of the data from figure 1. The TAFF pinning potential can be calculated with the slope of the linear parts. In order to find suitable fit parameters, it was assumed that the extrapolation of the slopes meet together in one common point at $1/T_c$, which is fixed from the $\rho(T,B = 0)$ data shown in figure 1.

Note that the smaller values occur due to much higher currents applied in the measurement. On the first sight, the weak pinning forces of the sample processed at 950 °C seem to be a contradiction to the results of Ref. [12], where magnetic measurements showed an increase of the pinning forces with increasing reaction temperature, but this gives important hints to the microstructure of the material.

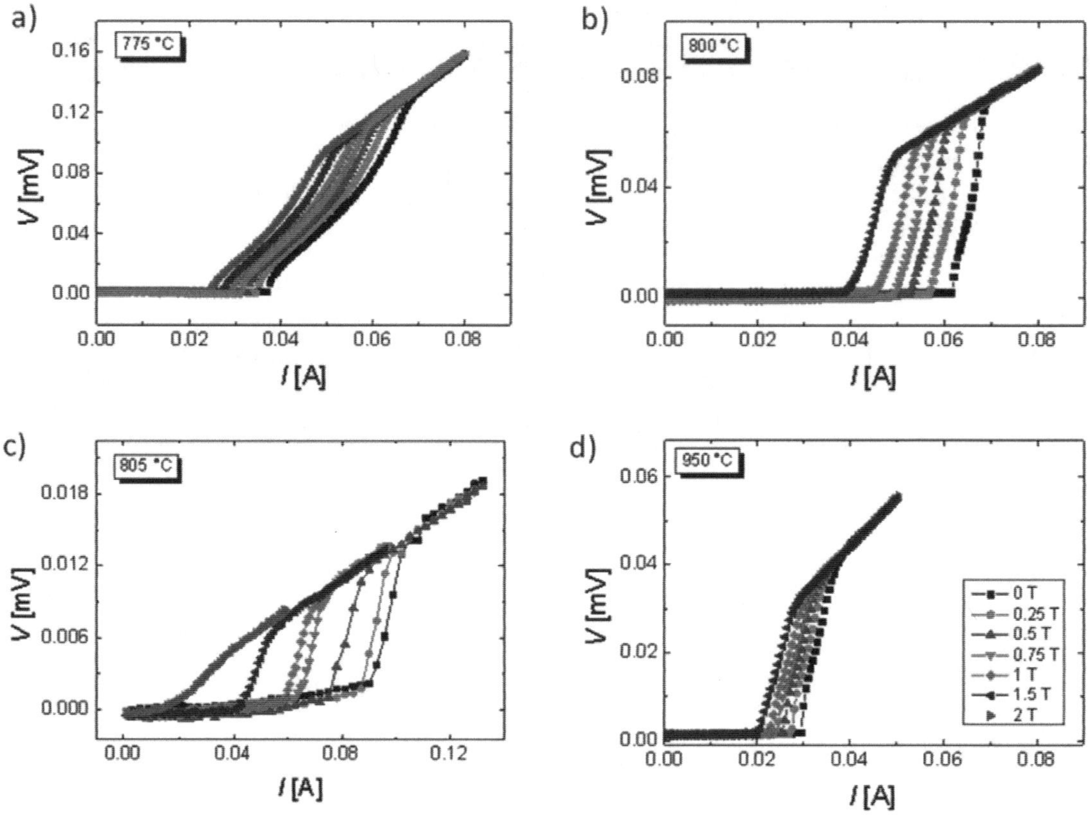

Figure 4. The voltage measured at 30 K as a function of current for different reaction temperatures in magnetic fields up to 2 T.

On increasing the reaction temperature, the number of point pinning centers, provided by MgB_4 inclusions, increases. But simultaneously the quality of the grain boundaries and therefore the intergrain connectivity decreases. Due to the different source of the shielding currents these currents are able to flow in single grains in the case of the magnetic measurements. Therefore, the sample sintered at 950 °C shows there a high pinning potential. In the electric measurements, the currents are forced to flow through the whole sample. This is the reason for the larger pinning forces of the samples sintered at around 805 °C in the case of the electric measurements due to a combination of strong point pinning centers provided by MgB_4 particles and a good intergrain connectivity.

Since the best process parameters for sintered MgB_2 samples were found at this point, one big disadvantage of these low cost samples remains. The material produced via the in-situ solid state reaction exhibits a density of just about 50% of the theoretically maximum density of MgB_2, which is low compared to other methods like spark plasma sintering [20]. This strongly requires an improvement of the mechanical properties of the resulting soft and porous samples. Therefore, silver-doped MgB_2 samples sintered at 775 °C (the optimum temperature found in Ref. [9]) were produced with silver contents ranging between 0 wt.-% and 10 wt.-%. In order to compare the electric and

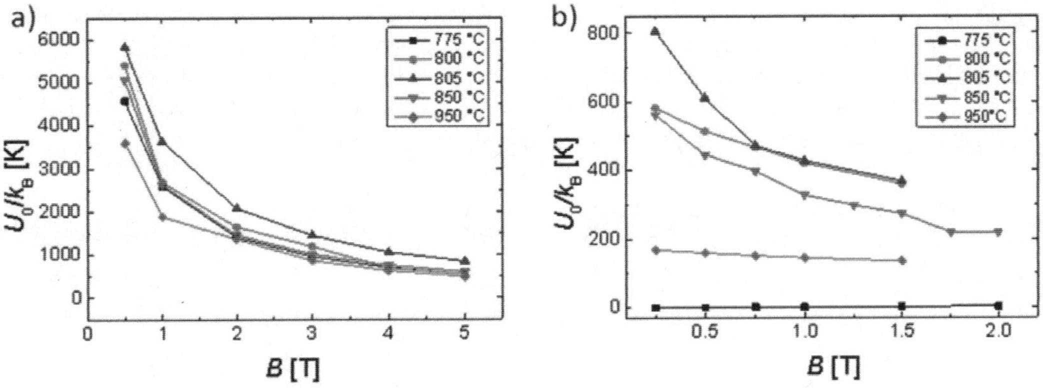

Figure 5. TAFF pinning potential (a) and flux creep pinning potential (b) as function of the magnetic field for various reaction temperatures at 30 K. In both cases, the sample sintered at 805 °C shows the highest value which indicates a combination of strong pinning centers and a good intergrain connectivity.

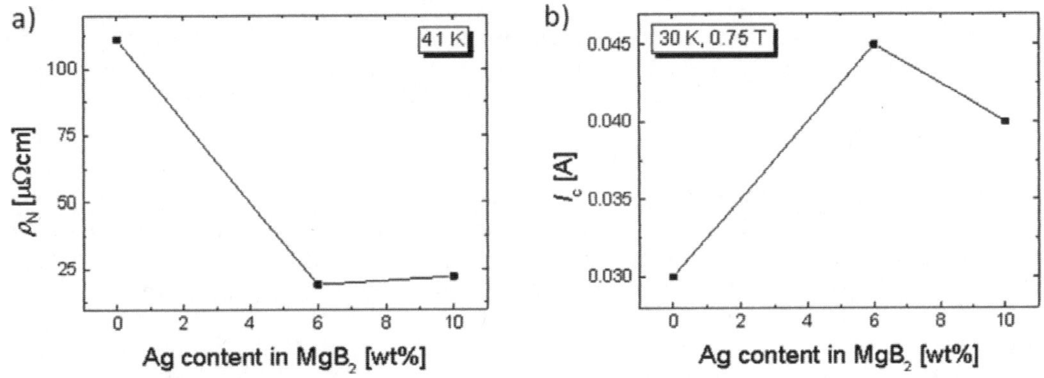

Figure 6. (a) Normal conducting electrical resistivity as function of Ag content taken at 41 K for the samples sintered at 775 °C. A decrease with silver doping is visible. (b) Critical current density as function of Ag content obtained at 0.75 T (30 K).

magnetic properties to the previous samples, the same transport measurements showed above were performed on these samples.

The electrical resistivities in the normal conducting state for the samples with silver contents of 0 wt.-%, 6 wt.-% and 10 wt.-%, which are obtained at 41 K, are shown in figure 6 (a). A decrease of the resistance with silver doping is clearly visible. The smallest value of 19 μΩcm is reached for an Ag content of 6 wt.-% of silver. Figure 6 (b) shows the critical current obtained at 0.75 T at 30 K. Higher values are reached with silver doping compared with the sample without silver. Also in this case, the highest value is reached for the sample with an Ag content of 6 wt.-%. The pinning potentials calculated from the resistivity and the *V-I* data are presented in figure 7. An increase of the TAFF pinning potential [figure 7 (a)] with silver doping is visible. This can be seen in the flux creep pinning potential as well. In both cases, the sample with an Ag content of 6 wt% shows the strongest pinning centers. These results can be explained with microstuctural observations in Ref. [11]. It was shown that silver doping leads to a formation of nanometer-sized $AgMg_3$ particles in the material, which can deal as additional pinning centers and hence increase the critical currents and improve the pinning properties. The study shows that the highest values for j_c is observed for Ag contents of 4-6 wt% due to a homogenous distribution of these particles, which is in agreement with our observations. This clearly demonstrates that silver doping leads to an improvement of the magnetic [11] and electric properties but also the mechanical properties can be improved, which is shown in Ref. [11].

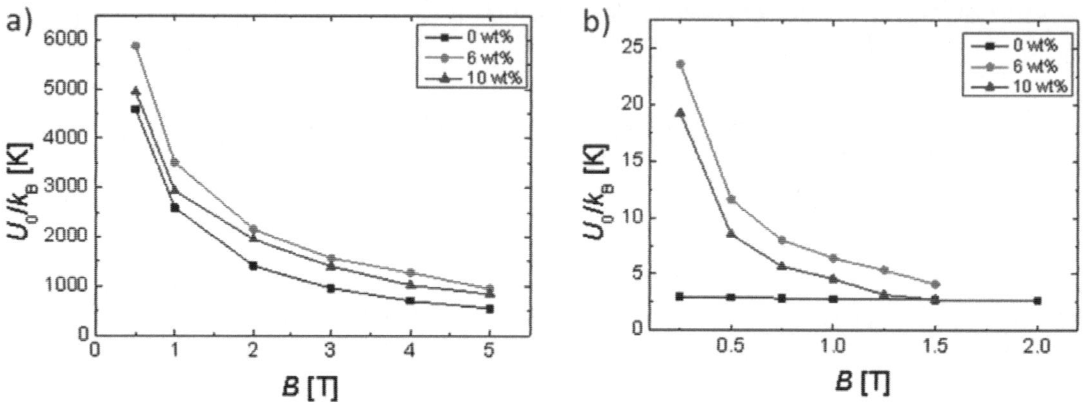

Figure 7. TAFF pinning potential (a) and flux creep pinning potential at 30 K (b) as function of the magnetic field for the different Ag contents. It is clearly visible that the silver addition improves the pinning centers in both cases. The highest potential is obtained at a silver content of 6 wt.-%.

Conclusions

To summarize, transport measurements were performed on various MgB_2 samples. The measurements reveal the best sintering temperature being around 805 °C for the low cost solid state reaction due a good intergrain connectivity and a high pinning potential. The properties of these samples can be improved further with silver doping which leads to stronger pinning, higher j_c values and higher trapped fields.

References

[1] Nagamatsu J, Nakagawa N, Muranaka T, Zenitani Y and Akimitsu J 2001 *Nature* **40** 1
[2] Moshchalkov V, Menghini M, Nishio T, Chen Q, Silhanek A, Dao V, Chibotaru L, Zhigadlo N and Karpinski J 2009 *Phys. Rev. Lett.* **102** 117001
[3] Muralidhar M, Inoue K, Koblischka M R, Tomita M and Murakami M 2014 *J. Alloys Compd.* **608** 102
[4] Yamamoto A, Ishihara A, Tomita M and Kishio K 2014 *Appl. Phys. Lett.* **105** 032601
[5] Tomsic M, Rindfleisch M, Yue J, McFadden K and Phillips J 2007 *J. Appl. Ceram. Technol.* **4** 3
[6] Buzea C and Yamashita T 2001 *Supercond. Sci. Technol.* **14** R115
[7] Glowacki B A, Majoros M, Vickers M, Evetts J E, Shi Y and McDougall I 2001 Supercond. Sci. Technol. **14** 193
[8] Naito T, Sasaki T and Fujishiro H 2012 *Supercond. Sci. Technol.* **25** 095012
[9] Muralidhar M, Ishihara A, Suzuki K, Fukumoto Y, Yamamoto Y and Tomita M 2013 *Physica C* **494** 85
[10] Muralidhar M, Inoue K, Koblischka M R, Tomita M and Murakami M 2014 *J. Alloy Compounds* **608** 102
[11] Muralidhar M, Inoue K, Koblischka M R, Murakami A and Murakami M 2015 *Adv. Eng. Mater.* **17** 831
[12] Koblischka M R, Wiederhold A, Muralidhar M, Inoue K, Hauet T, Douine B, Berger K, Murakami M and Hartmann U 2014 *IEEE Trans. Magn.* **50** 9000504

[13] Singh J P, Leu H J, Poeppel R B, Van Voorhees E, Goudey G T, Winsley K and Shi D 1989 *J. Appl. Phys.* **66** 3154
[14] Mendoza E, Puig T, Varesi E, Carrillo AE, Plain J and Obrados X 2000 *Physica C* **334** 7
[15] Wiederhold A, Koblischka M R, Inoue K, Muralidhar M, Berger K, Douine B, Hauet T, Murakami M and Hartmann U 2016, in: Superconductivity: Todays and Tomorrows Applications, ed. Muralidhar M, NOVA Science Pub., Commack, NY, p. 269
[16] Yamamoto A, Shimoyama J, Kishio K and Matsushita T 2007 *Supercond. Sci. Technol.* **20** 658
[17] Lei H, Wang K, Hu R, Ryu H, Abexkoon M, Bozin E S and Petrovic C 2012 *Sci. Technol. Adv. Mater.* **13** 054305
[18] Brandt E H 2012 *J. Nanosci. Nanotechnol.* **12** 1201
[19] Kobayashi H, Muralidhar M, Koblischka M R, Inoue K and Murakami M 2015 *Phys Procedia* **65** 73
[20] Noudem J G, Aburras M, Bernstein P, Chaud X, Muralidhar M and Murakami M 2014 *J. Appl. Phys.* **116** 163916

Application of shielding current in bulk HTS to control magnetic field distribution

T Kii
Institute of Advanced Energy, Kyoto University,
Gokasyo, Uji, Kyoto, Japan

E-mail: kii@iae.kyoto-u.ac.jp

Abstract. Superconducting shielding current is excited when external field is applied to superconductor. In case for field cooling of bulk superconductor, shielding current is an origin of strong trapped field. When external field is changed to a properly arranged bulk HTS array, various magnetic field distribution can be formed by an excited shielding current in each bulk HTS. This paper presents a simple intuitively method to design magnetic field distribution using supercurrents in bulk high-temperature superconductor (HTS) array. In this method, an ideal current path for intended field distribution is represented by shielding currents in bulk HTS array. Expected performance can be roughly estimated by using Biot-Savart law. As examples, Maxwell coil pair and helical field generator are designed. This method can be applied to design various magnet devices using bulk HTS array.

1. Introduction

Bulk high-temperature superconductors (HTS) have a large potential for various application because of its high current density. In case for rare-earth barium-copper-oxide, critical current density exceeds 10 kA/mm^2 below 10 K. This value is much higher than the equivalent current density of NdBFe permanent magnet; several hundred A/mm^2 or the engineering current density of superconducting wire; a few kA/mm^2. Moreover, minimum curvature of current path in bulk superconductor is not limited, because the current path is determined by the shape of bulk superconductor. Therefore, we focused on these two important properties of bulk superconductors, we have introduced a bulk HTS staggered array undulator which consists of stacked array of bulk HTS and a solenoid [1, 2] to generate strong periodic magnetic field for generation of synchrotron radiation, and successfully demonstrated stronger periodic magnetic field than the undulator which uses permanent magnets and superconducting wires [3]. A schematic drawing of the undulator is shown in figure 1. The periodically stacked bulk HTS of upper and lower line are shifted by half period and these are inserted in a solenoid. The rectangular hole along the central axis is a space for electron beam. Highest undulator field obtained with the undulator was 0.85 T for the period of 10 mm and the gap of 4 mm [3], and the performance was about 20% higher than the limit achieved with cooled combination of permanent magnet and ferromagnet. The experimental results are numerically evaluated using a simple loop current model [4] which is expressed using simple closed current paths along the shape of bulk HTS and an energy minimization method [5] which is based on energy minimizing treatment proposed by Badia et al [6]. The simple loop current model can reproduce magnetic saturation properties and the energy minimization method can quantitatively reproduce the experimental obtained magnetic field strength.

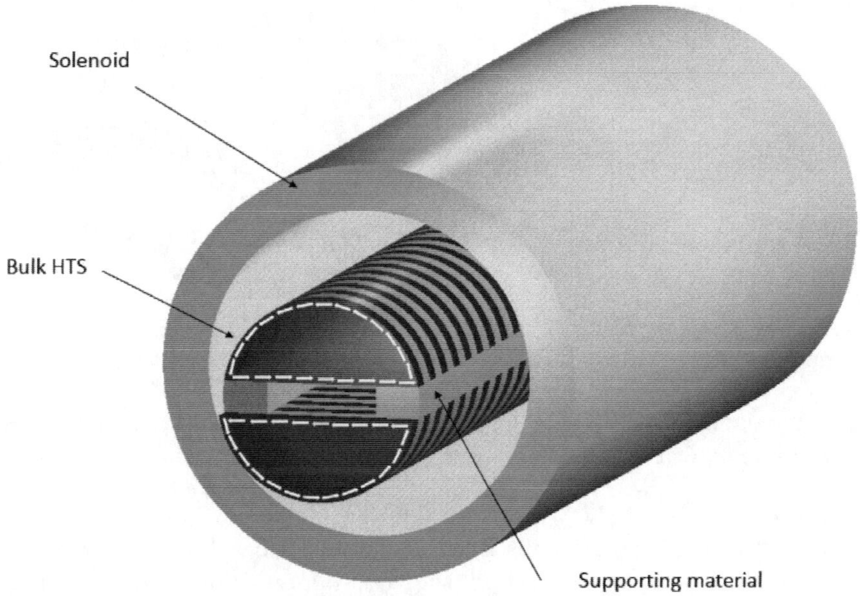

Figure 1. Schematic drawing of the bulk HTS SAU. When the external magnetic field is changed by the external solenoid, supercurrent is induced in each bulk HTS. Small arrows indicate the excited induced shielding supercurrents. Shielding currents of opposite direction are excited at above and below of the central axis. Transverse periodic magnetic field is generated on the central axis by the induced shielding supercurrents.

The intended periodic magnetic field distribution can be generated by superposition of the magnetic fi eld around the shielding current in the each bulk HTS. Most important feature of the operational sche me of the undulator is that the superconducting shielding currents exited in bulk HTS are highly desig ned to control magnetic field distribution. By generalizing this approach to other field distribution suc h as constant gradient field, helical field, bulk HTS array can be used to various applications and realiz e latent potential of bulk HTS.

2. Design procedure of bulk HTS array

In this section, a simple intuitively way to design HTS array is described without using numerical simulation tools. The approach using bulk HTS array and external field can be characterized by designing the current path which generates intended field. The simplest example of designed field is uniform magnetic field. An infinitely long solenoid or Helmholtz coil is used to generate uniform field. We can easily imagine the equivalent shape of the bulk HTS, which are infinitely stacked hollow disks and two hollow disks separated by a distance equal to the radius of the hollow disk respectively. Although the infinitely stacked hollow disks can realize better uniformity than the two hollow disks, use of the two hollow disks is adopted in the design procedure for generalization. An original hollow disk is represented by virtual current loop which consists of segmented fan shaped hollow parts as shown in figure 2.

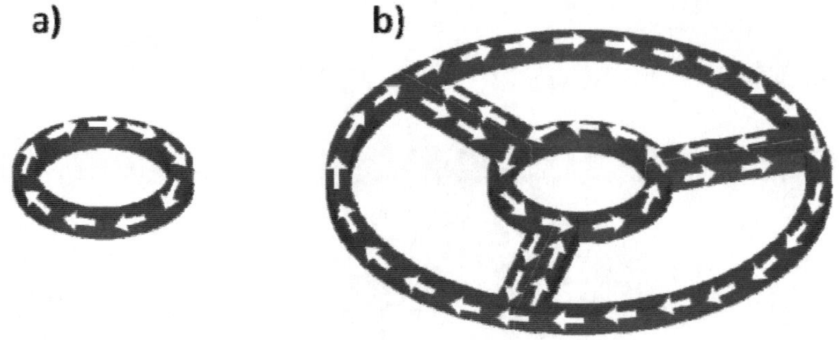

Figure 2. An alternative expression of hollow disk. In this example, an original ring (a) is represented using segmented three parts (b).

It is noted that the direction of excited shielding supercurrents is changed by segmentation, effect from straight lines cancels, and effect from outer circular current path can be reduced by increasing size.

2.1. Producing constant-gradient field

A Maxwell coil pair; two ring coils of radius R with opposite current flow are separated with distance $\sqrt{3}\,R$ can generate constant gradient field [7]. As shown in figure 3, this configuration can be realized by using combination of a hollow disk and a segmented current path which is introduced above. If the outer radius of the virtual current loop of the segmented parts is large enough, almost constant gradient field is generated between the rings. In order to obtain precise numerical estimation, mutual inductance among the hollow disk and segmented pars should be carefully considered and field distribution should be tuned by controlling precise shape, radius, and position of current paths, by using numerical tools such as an FEM code.

Figure 3. An example of Maxwell coil pair represented with hollow disk and segmented parts.

2.2. Producing helical field

Generation of strong helical magnetic field in short period is attractive for various applications such as spin physics and a helical undulator which generate circular polarized synchrotron radiation or photon beams carrying an orbital angular momentum [8]. The helical field can be produced using double helix as shown in figure 4.

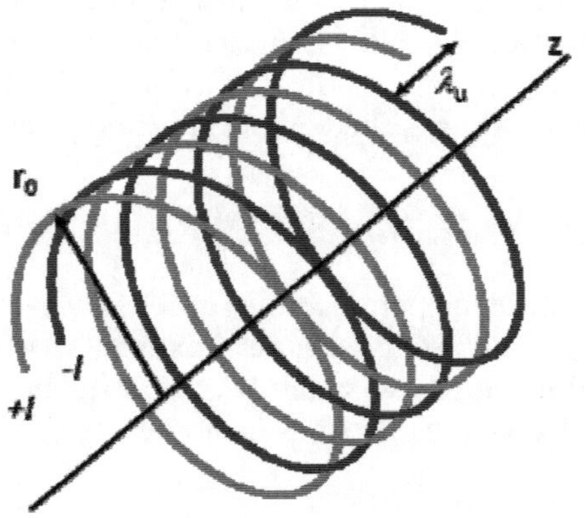

Figure 4. A schematic drawing of double helix.

Transverse magnetic field $B_\perp$ can be analytically expressed as follows [9]

$$B_\perp = \frac{\mu_0(+I-(-I))}{\lambda_u}\left\{\frac{2\pi r_0}{\lambda_u}K_0\left(\frac{2\pi r_0}{\lambda_u}\right)+K_1\left(\frac{2\pi r_0}{\lambda_u}\right)\right\} \quad (1).$$

Here, K_n is the modified Bessel function of the second kind. μ_0, λ_u, I, and r_0, are the permeability of free space, the period of spiral, the supercurrent, and the radius of helix, respectively. $+I$ and $-I$ represent currents with different direction in double helix. As same as the segmented hollow disk, one period of spiral current path can be expressed by segmented hollow parts as shown in figure 5.

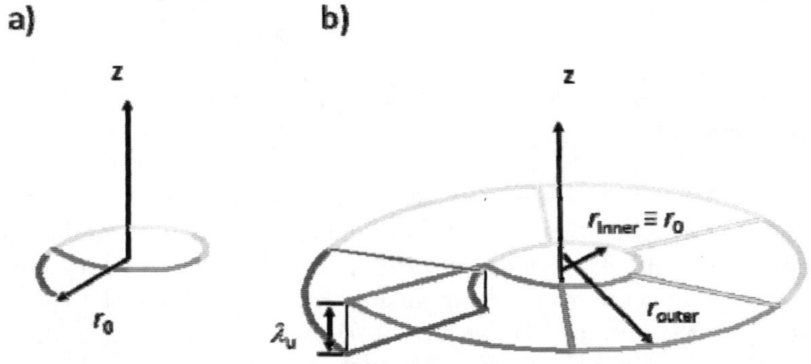

Figure 5. A schematic drawing of one period of spiral (a) and segmented parts (b). In this example, one period of spiral is divided into 6 parts.

Although the segmented outer virtual helix produce the opposite helical field to that of inner virtual loop, the opposite component rapidly decreases as increasing outer radius because the Bessel function decreases very quickly.

3. Estimation of magnetic field

In case of superconducting wire, amount of coil current is a variable which is directly controllable by a power supply. Hence we can easily estimate the current distribution by using Biot-Savart law. However, in case of bulk HTS array placed in external field, shielding current is not directly controlled by external field. In order to estimate required field strength to fill hollow parts with shielding currents, a very simple situation is considered. A superconducting ring with small curvature is placed in external magnetic field as shown in figure 6. As increasing the field strength, amount of shielding current increases and finally superconductor is filled with shielding current. At just before saturation, magnetic field at the small screened area is zero. It means that the external magnetic field strength required to fill the superconductor ring is same as the surface magnetic field strength with fully filled superconductor. As described above, the required magnetic field strength is roughly expressed as follows

$$B_{\text{external}} = B_{\text{surface}} = \frac{\mu_0 I_{\text{bulk}}}{2\pi r_{\text{bulk}}} \quad (2).$$

Here, μ_0, I_{bulk}, r_{bulk}, B_{external} and B_{surface} are the permeability of free space, total amount of shielding curr ent in the bulk ring, tube radius of the bulk ring, the applied external magnetic field, and the surface m agnetic field produced by the shielding current at saturation respectively. Under this condition, magnet ic field distribution can be easily calculated using Biot-Savart law.

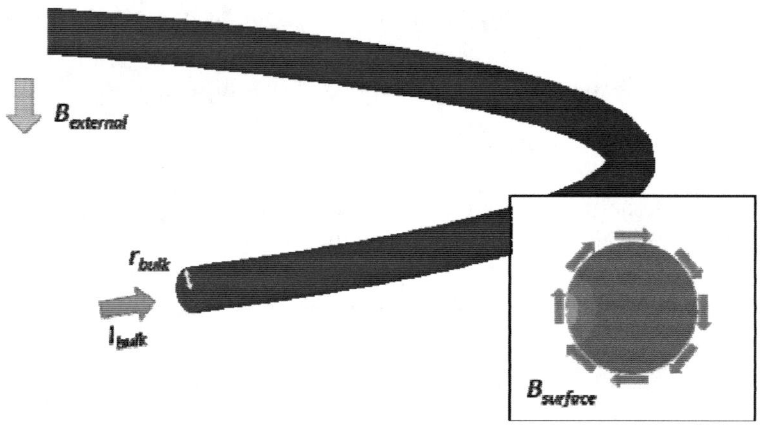

Figure 6. A schematic view of shielding current in the ring superconductor.

As already mentioned in previous section, detailed calculations are required to obtain precise field dist ribution. In order to clear advantage of the bulk array to superconducting wire, performance of helical field generator is compared numerically. Finite-size effect should be considered for actual device, i.e. current distribution, width, and thickness of the wire and bulk are considered. Here, we adopt effective radius of helix r_{eq} and equivalent current I_{eq} for the comparison. The equivalent radius is geometricall y averaged radius for finite size of coil. The equivalent current is simply defined by using width and th ickness of coils or hollow bulks, as $I_{\text{eq}} = J_c \times r_{\text{width}} \times D$. As an example of a double helix undulator, desig n parameter reported at advanced photon source in Argonne National Laboratory [10] is used. Schema tic drawing for the double helix is shown in figure 7.

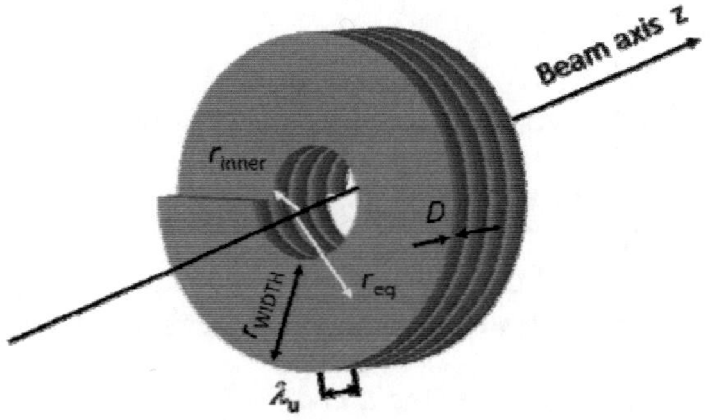

Figure 7. Geometrical model for double helix. Two spiral plates represent SC wires. Equivalent radius r_{eq} is defined as geometrical average radius.

Inner radius, coil thickness, width of LTS wire, pitch of the helix (period of spiral), and current density of low temperature superconducting (LTS) wire are 3.15 mm, 3.84 mm, 1.0 mm, 12.0 mm, and 1.0 k A/mm^2, respectively. Equivalent inner radius for double helix is 4.3 mm and the total current is 3.8 kA. The expected transverse helical field is 0.23 T. As for the HTS array, inner radius, outer radius, width of current region, pitch of helix, current density of HTS bulk are set to 3.15 mm, 17 mm, 3.84 mm, 1. 0 mm, 12.0 mm, and 10.0 kA/mm^2, respectively. The equivalent inner and outer radius of current loop for bulk HTS is 4.3 and 15 mm respectively. As the modified Bessel function in eq. (1) quickly decrea se as increasing radius, when the outer radius is 15 mm, the negative transverse field strength is less th an 0.6% of that of inner current path. When the operated temperature is set to around 10 K, current den sity of bulk HTS is about 10 kA/mm^2, thus the total current of 38 kA for bulk HTS array. By using this condition, strength of transverse magnetic field is calculated. The expected transverse helical field for the segmented bulk is 1.1 T for period of 12 mm and inner bore diameter of 6.3 mm. The transverse fi eld is about 5 times larger than that of the LTS wound wire undulator. Results are listed in table 1 toge ther with important parameters.

Table 1. Comparison of expected performance for the double helix LTS wound wire type undulator and the segmented single helix HTS bulk array type undulator

Type	Double Helix LTS Wound Wire	Segmented Single Helix HTS Bulk Array
Current Density J_c	1.0 kA/mm^2	10.0 kA/mm^2
Equivalent radius of helix r_{eq}	Inner 4.3 mm Outer ------	Inner 4.3 mm Outer 15 mm
Period of spiral λ_u	12 mm	12 mm
Thickness of coil or disk	1.0 mm	1.0 mm
Transverse field $B_\perp$	0.23 T	1.1 T

4. Conclusion

Bulk HTS have a high latent potential for magnetic application for those high critical current density and flexibility for current path. If we can properly arrange the bulk HTSs in the external field, various interesting strong magnetic field distribution can be produced. In this paper, basic concept for designing of bulk HTS array and several examples of bulk HTS array are introduced. We hope the new strategy for magnetic field distribution control using shielding current opens new research field using bulk superconductor.

Journal of Physics: Conference Series **695** (2016) 012005 doi:10.1088/1742-6596/695/1/012005

References:

[1] Kii T, et al. 2006 *Proc. 28th FEL Conf. FEL 2006*, **653**
[2] Kii T, et al. 2012 *IEEE Trans. on Applied Superconductivity Vol 22, No 3* **4100904**
[3] Kinjo R, et al. 2013 *Appl. Phys. Express vol. 6* **042701**
[4] Kinjo R, et al. 2009 Proc. 29th FEL Conf. FEL 2009, p 746,
[5] Kinjo R, et al. 2014 Phys. Rev. ST-AB 17 022401
[6] Badia A, Lopez C, and Giordano J L 1998 Phys. Rev. B 58, p 9440
[7] Clerk-Maxwell J 1873 *Treatise on Electricity and Magnetism* (Oxford: The Clarendon Press) p 319
[8] Bahrdt J, et al. 2013 *Phys. Rev. Lett. vol. 111* **034801**
[9] Smythe W R 1950 *Static and Dynamic Electricity* (McGraw-Hill. New York) p 277
[10] Kim S H 2012 *Magnetic Field Analysis of Helical Undulators* (ANL/APS/LS-331) p 1

Specific heat analysis of the low temperature anomalies in orthorhombic $PrBa_2Cu_3O_{6+x}$ ($x = 1$; $x = 0.95$) compounds

M Lahoubi
Badji-Mokhtar Annaba University, Faculty of Sciences, Department of Physics, Laboratory L.P.S., PO Box-12, 23000 Annaba, Algeria

E-mail: mlahoubi@hotmail.fr

Abstract. The specific heat $C_p(T)$ and entropy $S(T)$ properties of the orthorhombic $PrBa_2Cu_3O_{6+x}$ compounds in two states of oxygen concentration x, an over doped (OV) with $x = 1$ and an optimally doped (OP) with $x = 0.95$ are reanalyzed below the Néel temperature of the antiferromagnetic ordering of the Pr sublattice T_N = 17.5 and 14 K, respectively. Two simultaneous anomalies for both states are observed. The first one occurs near the previous spin reorientation phase transition temperature $T_2 \sim 11.5$ and ~ 9–10 K, respectively whereas the second one remains close to the so called low-critical temperature $T_{cr} \sim$ 4–5 K for the OV state as it has been reported before for the OP state. By fitting the $C_p(T)/T$ data to $A\{T^2\}^{-3/2} + \gamma + C\{T^2\}^1 + D\{T^2\}^2$ for $T < T_{cr}$ the four coefficients obtained with the best adjusted R-squared values are compared with previous findings. Reduced values for γ are confirmed in this work. The results which are well described by the contribution of the DT^5 term to $C_p(T)$ can be connected with the previous Pr–Cu(2) magnetic coupling that is sufficiently enough to cause a modest spin reorientation phase transition at T_2 and a critical magnetic behaviour below T_{cr}.

1. Introduction

It is well-known that $PrBa_2Cu_3O_{6+x}$ compounds in both orthorhombic and tetragonal phases exhibit absence of superconductivity which yields to anomalous magnetic and thermodynamic properties if the comparison is made with the high temperature $REBa_2Cu_3O_{6+x}$ (hereafter $REBCO_{6+x}$) superconductors containing other trivalent rare earth (RE) or yttrium ions [1-4]. In contrast to $YBCO_{6+x}$, for which the Cu(2) antiferromagnetism is lost once the level of oxygen concentration x exceeds 0.4, the cooperative antiferromagnetic (AFM) ordering of the Cu(2) sublattice lies at temperatures in the region of room temperature, throughout the entire range x = 0–1 with the Néel temperatures $T_N(x)$ = 250–350 K [1-3]. Another AFM ordering occurs additionally in the Pr sublattice within the temperature regime 1.5–30 K at the unexpectedly values of $T_N \sim$ 14–20 K for the optimally (OP) ($0.50 < x < 1$) and over (OV) ($x = 1$) doped states in the orthorhombic phase and ~ 7.5–13 K for the under doped (UN) ($0 < x < 0.50$) and oxygen-depleted (DE) ($x = 0$) states in the tetragonal phase. They are themselves anomalous and an order of magnitude higher than for the other RE ions sublattice in the $REBCO_{6+x}$ system for which T_N(RE) ~ 0–2.3 K.

Moreover, an anomalous feature of this $PrBCO_{6+x}$ antiferromagnet system is the existence of the Pr–Cu(2) magnetic coupling due to the pseudodipolar interactions discovered by Boothroyd *et al.* [5] on UN ($x = 0.35$) and OP ($x = 0.92$) single crystals which is responsible of the possible noncollinear ordering of both subsystems [6]. Meanwhile, the maximum near 13.75 K followed by a small minimum near 5 K with a rapid upturn up to 1.8 K in the (1/2, 12, 1) magnetic Bragg peak intensity originating from the order of the Pr sublattice was ascribed as resulting from the weak Pr–Cu(2) magnetic coupling in the neutron powder diffraction (NPD) data obtained on sample in the OP state with x = 0.93(2) [7] with T_N ~ 17 K. Another striking feature of the Pr sublattice concerns its spin reorientation phase transition (SRPT) firstly reported by Uma *et al.* [8] near T_2 ~ 13.4 K with a well marked anomaly in the magnetic specific heat $C_p^{magn}(T)/T$ curve and also around 13.5 K (in warming) in the neutron diffraction experiments carried out on single crystal in the OP state (x = 0.92) with T_N = 16.8 K, whereas the Cu(2) sublattice has been found weakly disturbed by a "modest" Pr–Cu(2) magnetic coupling. This SRPT at T_2 was confirmed later near 12 K in the $C_p(T)/T$ experiments and in the T-region 12–13.5 K in the derivative magnetization dM/dT measured in a magnetic field H = 0.1 T [9, 10] on OP and OV samples with T_N = 16.2 and 17.5 K, respectively and at T_2 = 12.5±0.5 K [11] on a single crystal in the OP state (x = 0.93) with T_N = 17.5±0.5 K in the time-of-flight neutron scattering measurements at high energy of the Cu spin excitations.

However, the more precise investigations performed by resonant X-ray scattering [12-15] on crystals in the OP state (x = 0.92) have revealed the existence of new transitions from commensurate (CO) to incommensurate (ICO) ordering of both Pr and Cu(2) sublattices in the 4.4–16.7 K temperature range. A first transition occurs from CO to ICO at $T_{CO\text{-}ICO}$ = 19 K which coincides with T_N = 17–20 K with the coexistence of a CO component of the bilayer Cu(2) ordering and an associated doubling of the magnetic unit cell in the ***c*** direction followed by a similar reorientation behaviour of the Pr sublattice from an ICO to a CO structure on lowering the temperature below T_2 ~ 9 K; whereas a second transition was observed when the ICO satellites vanished below $T_{ICO\text{-}CO}$ ~ 4.4 K i.e., near the previous and rarely reported low-temperature transition observed at T_{N2} ~ 5.2 K firstly by Jee *et al.* [16], Kebede *et al.* [17] and Li *et al.* [18] on $C_p(T)$ and $\chi(T)$ measurements for sample in the OV state.

In fact, the angular magnetoresistivity measurements performed by Sandu *et al.* [19] using a high magnetic field of 14 T applied to the CuO_2 planes of a $Y_{0.2}Pr_{0.8}BCO_{6+x}$ single crystal in the OP state have open the question of a great interest about the signature and the nature of this "5 K transition". These authors [19] have studied the interrelations between the Pr and Cu(2) sublattices as well as the crossovers and/or transitions between CO and ICO orders. The symmetry change of the ***c***-axis magnetoresistivity was ascribed to the AFM ordering of the Pr sublattice at T_N ~ 17 K and higher temperatures whereas another symmetry change of the *in*-plane magnetoresistivity at 6 K was attributed to the CO to ICO crossovers and/or transitions of both subsystems. The occurrence of this "5 K transition" has been also confirmed by the presence of a rather similar ferromagnetic peak (1/2, 1/2, 0) associated with the Pr ordering along the ***c*** axis [7] in the NPD experiments performed on the OP sample with x = 0.93(2) at 1.8 K in addition to the AFM magnetic Bragg peak (1/2, 1/2, 1/2) and by a sharp increase of the Grüneisen parameter below T_t ~ 5 K in the thermal expansion analyses performed on ceramic samples in the OP (x = 0.9) and UN (x = 0.4) states [20].

In the course of the previous thermodynamic investigations of $PrBCO_{6+x}$ in the orthorhombic phase the sensitivity of the electronic coefficient γ was highlighted firstly by Li *et al.* [18] who obtained a value of 114 mJ mol^{-1} K^{-2} for T < 5 K, 196 mJ mol^{-1} K^{-2} for 5 K < T < 17 K and an excess of 300 mJ mol^{-1} K^{-2} when extrapolated from T > 17 K by fitting $C_p(T)/T$ to $\gamma + CT^2 + DT^4$ and concluded that a reminiscent heavy fermions behavior must be commonly suggested. In their mixed $RE_{1-z}Pr_zBCO_{6+x}$ compounds with z > 0.55 and RE being a non magnetic ion Y, Eu or Lu [21-29] or in $PrB(Cu_{1-d}Ga_d)_3O_{6+x}$ with x = 0.9 [30] and $x \approx 1$ [31] while d = 0.04 and 0.08 [32], large values for γ confined between 102 and 300 mJ mol^{-1} K^{-2} were also reported. They differ significantly from many groups who used various fitting procedures and different fitting temperature ranges and/or fitting terms and then they led afterward to contradictory interpretations of the $C_p(T)$ data. Based on our recent study of the thermodynamic properties of non-superconducting $PrBCO_{6+x}$ ceramics in the tetragonal phase [33] in the UN (x = 0.44) and DE (x = 0) states, two transitions were evidenced below T_N ~ 9 and 10 K, respectively near T_2 ~ 6–7 K and T_{cr} ~ 4–5 K labelled as the "low-critical temperature". However, the fitting results obtained in the low-temperature region $T < T_{cr}$ using a new $C_p(T)/T$ versus T^2 development taking into account the $A\{T^2\}^{-3/2}$ term associated to the hyperfine nuclear contribution have led to some considerable reduced values of γ, 45.6–59.6 mJ mol^{-1} K^{-2} by comparison with those

Journal of Physics: Conference Series **695** (2016) 012006 doi:10.1088/1742-6596/695/1/012006

found on our first work made previously on a ceramic sample in the OP state (x = 0.95) [34]. We believe that the unresolved question concerning the origin [27] and/or the consistency of the large values reported for γ in the orthorhombic $PrBCO_{6+x}$ system needs to be clarified.

2. Experimental

The polycrystalline samples prepared using the standard ceramic method were synthesized originally at Orsay Laboratory, Paris, France and Institute of Physics of ASCR, Praha, Czech Republic for the OV (x = 1) and OP (x = 0.95) states, respectively [33-35 and refs. therein]. The magnetic susceptibility $\chi(T) = M/H$ data under a field H = 100 Oe in the 5–300 K temperature range obtained for the sample in the OV state with a weight of 47.6 mg, were fitted above T_N = 17.8 K to a modified Curie Weiss law $\chi(T) = \chi_0 + C/(T-\theta_p)$. Thus, the paramagnetic temperature $\theta_p \approx -7.2$ K, the effective moment of the Pr ion μ_{eff} = 2.63 μ_B/f.u. and the Curie constant $C = 1.21\times10^{-3}$ cm^3/g K are found in good agreement with the literature [21], [27]. Whereas for the OP state and for samples in the tetragonal phase UN (x = 0.44) and DE (x = 0) some structural and magnetic characterizations were reported elsewhere [33], [34]. Here we report in figures 1and 2 the $C_p(T)$ and entropy $S(T)$ data [33-35] re-plotted in the 1.4–25 K temperature range. It is not surprising that no obvious anomaly can be seen below T_N whatever x.

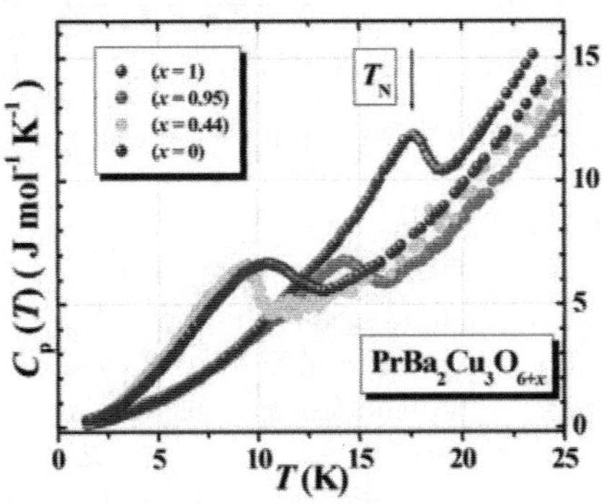

Figure 1. $C_p(T)$ vs T for $PrBCO_{6+x}$ samples in the OV, OP, UN and DE states [33-35].

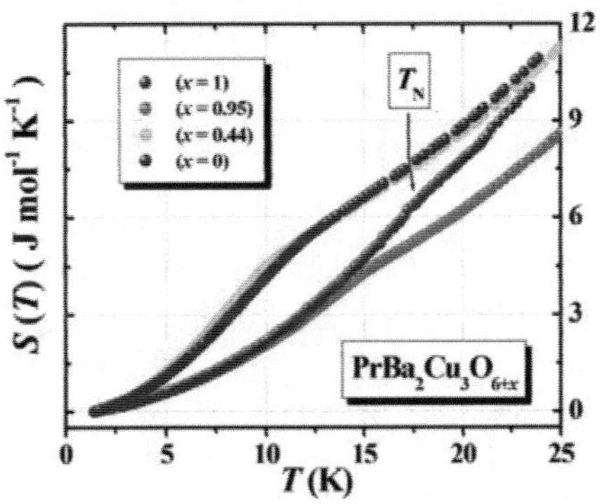

Figure 2. $S(T)$ vs T for $PrBCO_{6+x}$ samples in the OV, OP, UN and DE states [33-35].

3. Results and discussion

3.1. Low temperature anomalies

Shown in figure 3 is the $C_p(T)/T$ versus T in the 1.4–25 K temperature range for the OV state (x = 1).

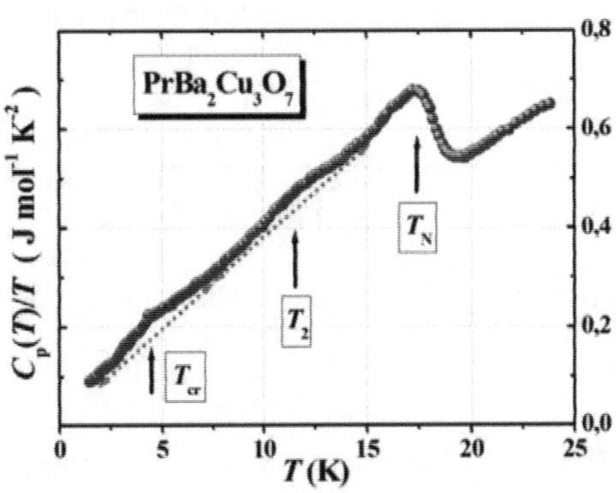

Figure 3. $C_p(T)/T$ vs T for the OV sample.

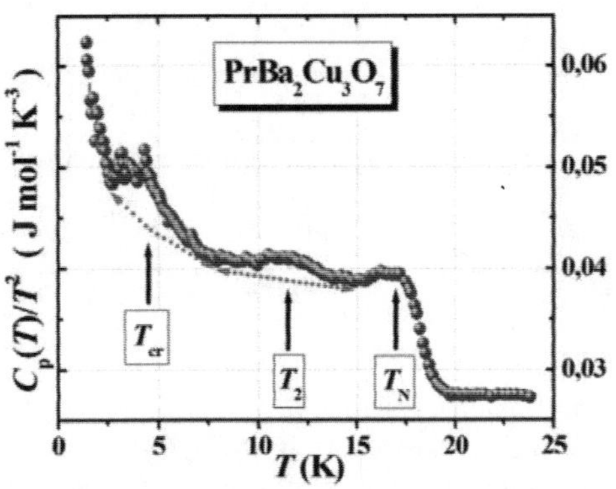

Figure 4. $C_p(T)/T^2$ vs T for the OV sample.

The curve develops a characteristic broad shoulder with a maximum at 17.5 K which coincides with T_N in agreement with the values reported in previous works for samples in the OV states [21], [27]. However, one can observe two extended "humps" centred near ~ 11.5 and ~ 4.5 K and expanded in the T-region of width 7.5–15 K and 1.4–7.5 K, respectively. These anomalies labeled T_2 and T_{cr} respectively, appear simultaneously with more evidence in the $C_p(T)/T^2$ versus T curve plotted in figure 4 where a well marked peak near T_{cr} is accompanied just below 2.5 K by a rapid upturn.

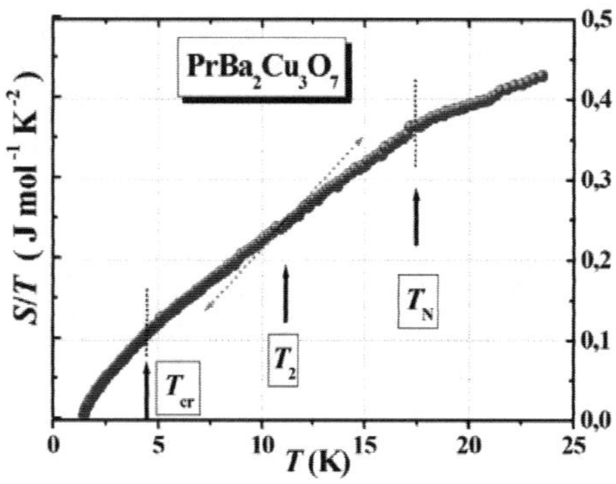

Figure 5. $S(T)/T$ vs T for the OV sample.

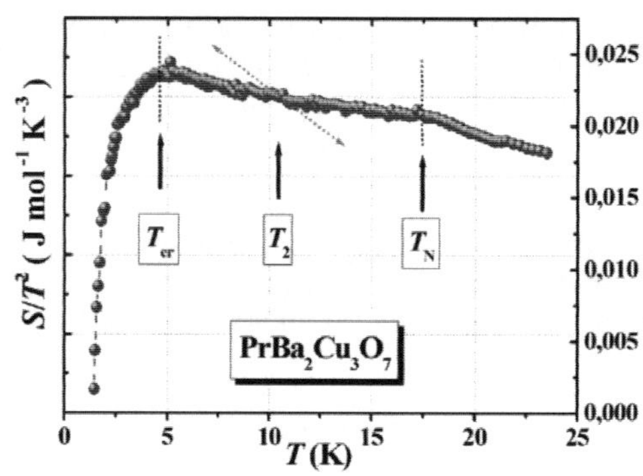

Figure 6. $S(T)/T^2$ vs T for the OV sample.

Figures 5 and 6 exhibit the temperature dependence of the entropy $S(T)$ in the no customary representations $S(T)/T$ and $S(T)/T^2$ in order to emphasize well every anomalies which are not clearly detected in the $S(T)$ versus T. Both curves present explicitly two different behaviors through T_{cr} and T_N with a small inflexion point near T_2 in the $S(T)/T^2$ curve where a maximum value of 0.024 mJ mol^{-1} K^{-3} is attained at T_{cr} which is followed by the onset of a rapid decrease up to 1.4 K. For a better visualization below T_N of the signature of these anomalies, the plots of $\Delta C_p(Pr)/C_p(PrBCO_{6+x})$ and $C_p(PrBCO_{6+x})/C_p(YBCO_7)$ versus T are presented in figures 7 and 8 for the OV and OP states, respectively. The features observed in the curves confirm that transitions occur near T_2 and at T_{cr}.

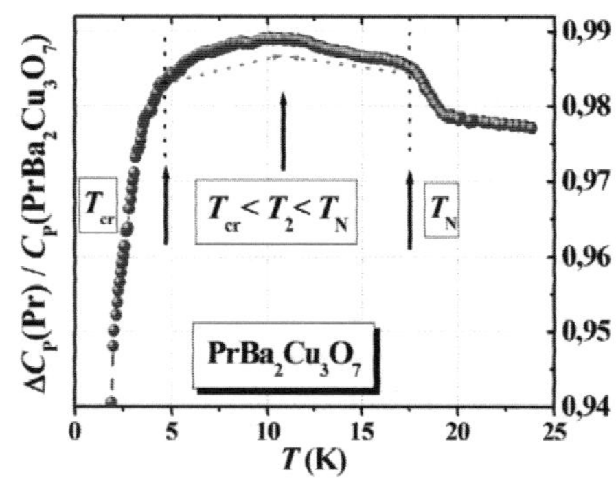

Figure 7. $\Delta C_p(Pr)/C_p(PrBCO_{6+x})$ vs T for the OV sample.

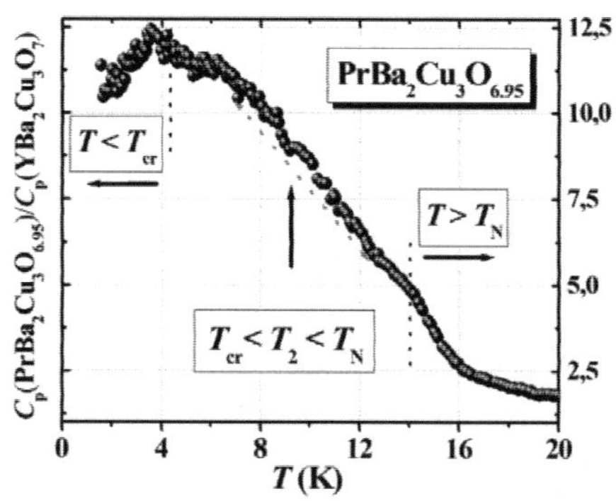

Figure 8. $C_p(PrBCO_{6+x})/C_p(YBCO_7)$ vs T for the OP sample.

3.2. Phenomenological Analysis

We subtracted the $C_p(T)$ of $YBCO_7$ in the OV state [27] given in the same temperature range and considered as the "background" contribution to the $C_p(T)$ data to determine $\Delta C_p(T)$, which is assumed in a good approximation to characterize the magnetic and electronic $C_p(T)$ contributions due to Pr only. According to our recent works [33, 34], the $C_p(T)$ data versus T can be described in a good approximation using the following expansion based on the sum of four terms:

$$C_p(T) = AT^{-2} + \gamma T + CT^3 + DT^5, \qquad (1)$$

where the first term is the hyperfine nuclear contribution caused from the hyperfine-field and/or electric quadrupolar interactions while the second term is the electronic linear contribution. The third and the four terms describe two different contributions where the coefficients are respectively the sums $C = \beta + M$ and $D = \delta + m$. The first one concerns the contributions of the harmonic (Debye) phonon of the lattice part (coefficients β) and the three dimensional AFM magnons which result from the Pr–Pr and Cu(2)–Cu(2) interactions (coefficient M); the second one characterizes the anharmonic phonon part of the lattice (coefficient δ) and the non linear effects (coefficient m) which are assumed due to the unconventional Pr–Cu(2) magnetic coupling between the Pr and Cu(2) sublattices [5, 6] in agreement with the appearance of the SRPT observed at T_2 [8-10]. In this case, the $C_p(T)/T$ data versus $\{T^2\}$ can be written according to the following development

$$C_p(T)/T = A\{T^2\}^{-3/2} + \gamma + C\{T^2\}^1 + D\{T^2\}^2, \qquad (2)$$

where the fitting T-range is now well defined as $T \le T_{cr}$. All least-squares fitting coefficients, A, γ, C and D are gathered on table 1 with the corresponding values of the adjusted-R squared coefficient.

Table 1. Fitting coefficients, A, γ, C, and D from the curves $C_p(T)/T$ versus T^2 of $PrBCO_{6+x}$ with ($x = 1$, OV) and ($x = 0.95$, OP) for $T \le T_{cr}$ with a comparison with previous results in OV, OP, UN, DE states and on $YBCO_7$.

State x	T_N (K)	A (mJmol^{-1}K)	γ (mJmol^{-1}K^{-2})	$C = \beta + M$ (mJmol^{-1}K^{-4})	$D = \delta + m$ (mJmol^{-1}K^{-6})	R (%)	T-range (K)
1	17.5	21.46	60.39	11.43	− 0.17(40)	99.58	1.4–4.5
1	17.5	8.01	65.47	10.64	− 0.14(89)	99.34	1.4–4.0
0.95	14.0	17.20	91.88	7.98	− 0.12(42)	99.36	1.4–5.0
0.95	14.0	21.08	90.34	8.24	− 0.13(44)	99.20	1.4–4.5
1 [a]	17.0	.	114	.	.	.	1.5–5.0
0.95 [b]	17.0	61.00	102	6.99	.	.	0.3–1.5
0.95 [c]	14.0	60.00	80.0	8.50	$+ 1.00\times10^{-3}$	99.50	1.4–2.5
0.90 [d]	17.0	44.00	118	5.50	.	.	0.6–5.0
0.90 [e]	14.0	93.00	146	8.30	.	.	0.6–5.0
0.90 [f]	10.0	77.00	168	11.40	.	.	0.6–5.0
0.44 [g]	9.0	60.00	59.6	32.00	− 0.54	99.40	1.4–5.0
0.00 [h]	10.0	217.0	45.6	30.00	− 0.68	99.00	1.4–4.0
$YBCO_7$ [i]		7.0	5.1	0.290	$+ 0.25\times10^{-3}$	.	1.5–10
$YBCO_7$ [j]		61.0	7.0	0.230	$+ 1.00\times10^{-3}$	.	1.5–17

[a] $PrBCO_{6+x}$ Li *et al.* [18].

[b] $PrBCO_{6+x}$ Phillips *et al.* [24].

[c] $PrBCO_{6+x}$ Younsi *et al.* [34].

[d] $PrBCO_{6+x}$; [e] $PrB(Cu_{1-d}Ga_{d=0.04})_3O_{6+x}$; [f] $PrB(Cu_{1-d}Ga_{d=0.08})_3O_{6+x}$ Yang *et al.* [30, 31]. The orthorhombic-tetragonal transition was observed at $d = 0.075$, Yang and Lin [32].

[g] $PrBCO_{6+x}$ ($x = 0.44$); [h] $PrBCO_{6+x}$ ($x = 0$) Younsi *et al.* [33].

[i] $YBCO_7$ Sasaki *et al.* [36].

[j] $YBCO_7$ Hilscher *et al.* [27].

Unfortunately, our comparison with others findings (see the notes on table 1) can not be complete owing the fact that their adjusted-R determinations were not indicated.

Shown in figures 7 and 8 are the best fitting least-squares of the $C_p(T)/T$ versus T^2 obtained for the OV (x = 1) and OP (x = 0.95) states where their adjusted-R squared coefficients are 99.58% and 99.36%, respectively. It is found that the best fitting γ values, respectively, 60.39 and 91.88 mJ mol^{-1} K^{-2} are significantly lower by about 10–40% and 22–49% than those obtained preferentially by Phillips *et al.* [24] and Yang *et al.* [30, 31] respectively.

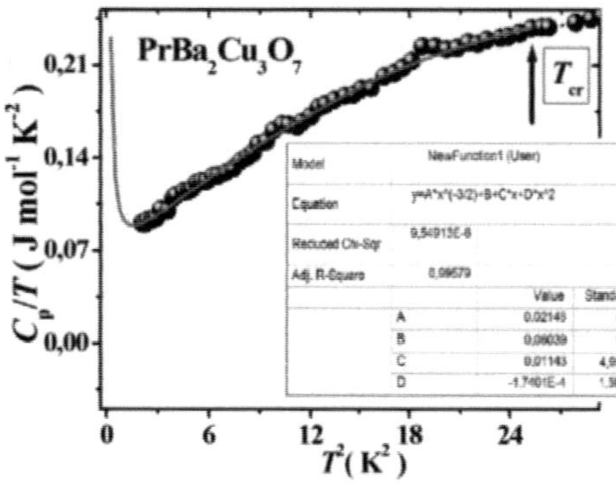

Figure 7. Best fit of the data in the OV state (x = 1) by a $C_p(T)/T$ vs T^2 curve for $T < T_{cr}$.

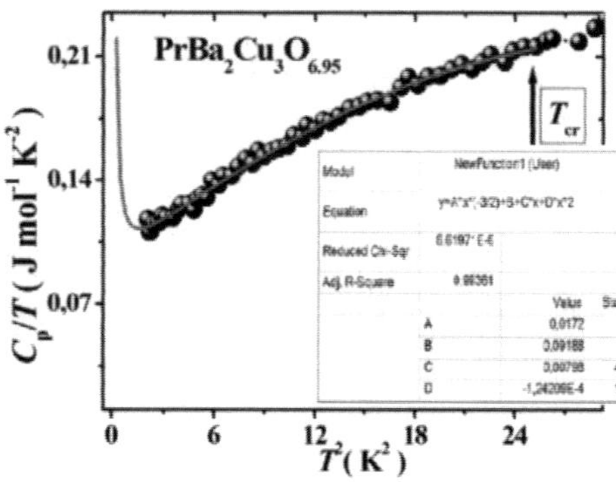

Figure 8. Best fit of the data in the OP state (x = 0.95) by a $C_p(T)/T$ vs T^2 curve for $T < T_{cr}$.

These two groups of authors have used the same development of equation (2) for $C_p(T)/T$ versus T^2 without the DT^5 term and have taken as the fitting range of temperatures 0.3–1.5 and 0.6–5.0 K respectively, although the anomaly at T_{cr} not being observed. These reduced values are comparable or slightly higher by a factor two to the fullest than those obtained for the UN (x = 0.44) and DE (x = 0) states, 59.6 and 45.6 mJ mol^{-1} K^{-2}, respectively [33] in agreement with the insulating character of these non-superconducting materials.

Our C fitting values, 11.43 and 7.98 mJ mol^{-1} K^{-4}, are rather comparable with earlier reports on OP state [24, 34] and confirm that the existence of a T^3 term is associated mainly with the three dimensional AFM ordered states of both Pr and Cu(2) sublattices, if the comparison is made in a first approximation with the corresponding values 0.23–0.29 mJ mol^{-1} K^{-4} of the alone parameter β in $YBCO_7$ [27, 36]. In contrast, they are significantly lower, in a ratio of order ¼ by comparison with those reported for the UN (x = 0.44) and DE (x = 0) states 32 and 30 mJ mol^{-1} K^{-4}, respectively [33], which is consistent with the conclusion that oxygen depletion induces more insulating behaviour with a strong effect on the magnetic contribution.

The D fitting values, –0.17 and –0.12 mJ mol^{-1} K^{-6} obtained for the OV and OP states respectively, have a negative sign and a high magnitude. These results could be unexpected if the comparison is made with our previous determination $+1.00\times10^{-3}$ mJ mol^{-1} K^{-6} obtained on the same OP state by taking in the procedure a small fitting temperature range (1.4–2.5 K) [34]. But in fact, they are lower in magnitude –0.54 and –0.68 mJ mol^{-1} K^{-6} than those found for the UN (x = 0.44) and DE (x = 0) states, respectively [33]. With some residual phonon anharmonicity contribution of the lattice (see the notes on table 1), these meaningful determinations of the coefficient D imply that these non linear effects associated with the Pr–Cu(2) magnetic coupling [5] are obviously much important, in agreement with the theoretical studies based on the presence of pseudodipolar and quadrupolar interactions in this $PrBCO_{6+x}$ system whatever the phase, orthorhombic or tetragonal [37, 38].

4. Conclusion

The thermodynamical properties of two non-superconducting $PrBCO_{6+x}$ compounds in two oxygen concentrations, an OV (x = 1) and OP (x = 0.95) states have been revisited by analyzing the specific heat $C_p(T)$ and entropy $S(T)$ data at low temperatures. The first important feature revealed in this study is the existence of two simultaneous anomalies below T_N, the Néel temperature of the antiferromagnetic ordering of the Pr sublattice which is located at 17.5 and 14 K, respectively. Using the temperature dependencies of the following representations for $C_p(T)$ and $S(T)$, $C_p(T)/T$, $C_p(T)/T^2$, $S(T)/T$ and $S(T)/T^2$ the first anomaly is fairly observed at $T_2 \sim 11.5$ and ~ 9–10 K for the OV and OP states, respectively. Whereas the second anomaly is well marked at the low-critical temperature $T_{cr} \sim$ 4–5 K for the OV state as it has been reported before for the OP state. From the fitting of the $C_p(T)/T$ data versus T^2 below T_{cr}, reduced values of γ are obtained in comparison with the overestimated determinations deduced before. A possible explanation of these results well described by the existence of T^5 term can be connected with the Pr–Cu(2) magnetic coupling of pseudodipolar symmetry that is sufficiently enough to cause a modest SRPT at T_2 and a critical magnetic behaviour below T_{cr}. More detailed studies are needed to confirm that the Pr sublattice plays a main role while that of the Cu(2) subsystem seems less significant in this unusual magnetic reordering below T_N. So, the same phenomenological analysis of the magnetic and electronic contributions due to the Pr ions only to $C_p(T)$ may be considered as a possible extension of this work to ascertain these conclusions.

5. References

[1] Lynn J W, Rosov N, Barilo S N, Kurnevitch L and Zhokhov A 2000 *Chinese J. Phys.* **38** 286

[2] Radousky H B 2000 Magnetism in the Pr-Containing Cuprates *Series in Modern Condensed Matter Physics* vol 11, Magnetism in Heavy Fermion Systems ed H B Radousky (World Scientific Publishing Co Pte Ltd) chapter 6 p 345

[3] Lynn J W and Skanthakumar S 2001 Neutron Scattering Studies of Lanthanide Magnetic Ordering *Handbook on the Physics and Chemistry of Rare Earths* vol 31, ed K A Gschneidner Jr, L Eyring and M B Maple (Amsterdam: Elsevier) chapter 199 pp 313–348

[4] Akhavan M 2002 *Physica* B **321** 265

[5] Boothroyd A T, Longmore A, Andersen N H, Brecht E and Wolf Th 1997 *Phys. Rev. Lett.* **78** 130

[6] Boothroyd A T 1998 *Physica* B **241–243** 792

[7] Skanthakumar S, Lynn J W, Rosov N, Cao G and Crow J E 1997 *Phys. Rev.* B **55** R3406

[8] Uma S, Schnelle W, Gmelin E, Rangarajan G, Skanthakumar S, Lynn J W, Walter R, Lorenz T,. Büchner B, Walker E and Erb A 1998 *J. Phys.: Condens. Matter* **10** L33

[9] Levy G, Maiorov B, Corvalán M S, Fainstein A and Nieva G 2002 *Physica* B **320** 333

[10] Levy G 2001 Master Thesis in Spanish unpublished *Estudio de las propiedades estructurales, superconductoras y magneticas en la solución sólida* $Y_{1-x}Pr_xBa_2Cu_3O_y$ *con diferentes concentraciones de oxígeno* Universidad Nacional de Cuyo, Instituto Balseiro, Argentina, pp 1–80

[11] Boothroyd A T, Andersen N H, Larsen B H, Zhokhov A A, Frost C D and Adroja D T 2005 *Phys. Rev.* B **71** 094514

[12] Hill J P, Boothroyd A T, Andersen N H, Brecht E and Wolf Th 1998 *Phys. Rev.* B **58** 11211

[13] Boothroyd AT, Hill J P, McMorrow D F, Andersen N H, Stunault A, Vettier C and Wolf Th 1999 *Physica* C **317–318** 292

[14] Hill J P, Mc Morrow D F, Boothroyd A T, Stunault A, Vettier C, Berman L E, Zimmermann M v and Wolf Th 2000 *Phys. Rev. B* **61** 1251.

[15] Boothroyd A T 2000 *J. Alloy. Compd.* **303–304** 489

[16] Jee C S, Kebede A, Yuen T, Bloom S H, Kuric M V, Crow J E, Guertin R P, Mihalisin T, Myer G H and Schlottmann P 1988 *J. Magn. Magn. Mater.* **76–77** 617

[17] Kebede A, Jee C S, Schwegler J, Crow J E, Mihalisin T, Myer G H, Salomon R E, Schlottmann P, Kuric M V, Bloom S H and Guertin R P 1989 *Phys. Rev.* B **40** 4453

[18] Li W-H, Lynn J W, Skanthakumar S, Clinton T W, Kebede A, Jee C-S, Crow J E and Mihalisin T 1989 *Phys. Rev.* B **40** 5300

[19] Sandu V, Zhang C, Almasan C C, Taylor B J and Maple M B, 2006 *J. Phys.: Conf. Ser.* **51** 231

[20] Hean P, Lahoubi M, Harat A, Fillion G, Schmitt D, Bioud H and Hejmánek J 2000 *Physica* B **284–288** 1035
[21] Felner I, Yaron U, Nowik I, Bauminger E R, Wolfus Y, Yacoby E R, Hilscher G and Pillmayr N 1989 *Phys. Rev* B **40** 6739
[22] Maple M B, Ferreira J M, Hake R R, Lee B W, Neumeier J J, Seaman C L, Yang K N and Zhou H 1989 *J. Less. Common. Met.* **149** 405
[23] Ghamaty S, Lee B W, Neumeier J J, Nieva G and Maple M B 1991 *Phys. Rev.* B **43** 5430
[24] Phillips N E, Fisher R A, Caspary R, Amato A, Radousky H B, Peng J L, Zhang L and Shelton R N 1991 *Phys. Rev.* B **43** 11488
[25] Nieva G, Ghamaty S, Lee B W, Maple M B and Schuller I K 1991 *Phys. Rev.* B **44** 6999
[26] Schaudy G, Holubar T, Pillmayr N, Leitenbauer G, Hilscher G, Rogl P and Hanninger A 1992 *J. Magn. Magn. Mater.* **104–107** 477
[27] Hilscher G, Holland-Moritz E, Holubar T, Jostarndt H-D, Nekvasil V, Schaudy G, Walter U and Fillion G 1994 *Phys. Rev.* B **49** 535
[28] Fisher R A, Phillips N E, Wright D A, Radousky H B, Goodwin T J, Peng J L, Shelton R N and Zhang L 1994 *Physica* C **235–240** 1749
[29] Ho J C, Wu CY, Kumar K, Wu M K, Guan W Y and Xu Y H 1994 *Physica* C **229** 351
[30] Yang H D, Tsay H L, Shih C R, Meen T H and Chen Y C 1995 *Phys. Rev.* B **52** 15099
[31] Yang H D, Tsay H L, Juang F L, Wang W M, Huang W J, Chen C M and Meen T H 1996 *Chinese J. Phys.* **34** 252
[32] Yang H D and Lin M W 1991 *Phys. Rev.* B **44** 5384
[33] Younsi W, Lahoubi M and Soltani M-L 2012 *J. Low Temp. Phys.* **166** 218
[34] Younsi W, Lahoubi M, Soltani M-L, Schmitt D and Hejtmánek J 2009 *J. Phys.: Conf. Ser.* **150** 052290
[35] Younsi W 2005 Magister Thesis in French unpublished *Contribution à l'étude des propriétés magnétiques et électroniques des composés* $PrBa_2Cu_3O_{6+x}$ *(x = 0, 0.44, 0.95, 1)* Badji-Mokhtar Annaba University, Algeria pp 1–58
[36] Sasaki T, Kobayashi N, Nakatsu O, Matsuhira T, Tokiwa A, Kikuchi M, Syono Y and Muto Y 1988 *Physica* C **153-155**, Part 2, 1012
[37] Maleev S V 1998 *JETP Lett.* **67** 947
[38] Kiss A and Simon F 2010 *Phys. Rev.* B **82** 174413

Acknowledgments
This work was supported in part by the Algerian Ministry of Higher Education and Scientific Research under the CNEPRU Project (Grant no. D01120060130) and within a previous collaboration between Badji-Mokhtar Annaba University (Algeria) and Joseph Fourier University of Grenoble-UJF (France). The author wishes to acknowledge J. Hejtmánek (Institute of Physics of ASCR, Praha, Czech Republic) for providing the initial PBCO specimen and G. Hilscher (Institut für Experimentalphysik, Technische Universität Wien, Vienna, Austria) for his complete support some years ago for this special study. Special acknowledgment to P. M. Grant, Ph.D., Physicist and Science Writer (w2agz Technologies, San Jose California, USA) who encouraged me during the last EUCAS 2015 conference to continue the research on these compounds.

Local flux intrusion in HTS annuli during pulsed field magnetization

V S Korotkov[1], E P Krasnoperov[1,2] and A A Kartamyshev[1]

[1] National Research Center "Kurchatov Institute", Kurchatov Square 1, 123182, Moscow, Russia
[2] Moscow Institute of Physics and Technology, Institutskiy Pereulok, 9, 141700, Dolgoprudniy, Russia

E-mail : vasmephi@mail.ru

Abstract. During pulsed field magnetization of melt-grown HTS flux jumps can occur and the shielding current falls by 10-20 times. As the duration of pulse is shorter than the temperature relaxation time (<< 1 s), the circular current remains small during the field falling. The residual trapped field in the hole of the annulus has a direction opposite to that of the pulsed field. Small circular current and high critical current density are explained by the fact that flux moves through narrow regions of the annulus body. The angle of the sector with "soft flux" (i.e. a low J_c region) is estimated to be ~ 7 deg.

1. Introduction

Flux jumps in hard superconductors are extensively studied. At low temperatures in the conditions of the slow magnetization (the isothermal limit) flux jumps arise due to the thermo-magnetic instability [1]. High-temperature superconductors (HTS) have high heat capacity and the thermo-magnetic instability is usually not observed at $T > 40$ K [2]. At high temperature flux jumps arise in conditions of large disturbances, in particular, fast variation of the magnetic field. The study of flux jumps in HTS during pulsed field magnetization (PFM) is of interest because this process is useful for industrial and power applications.

2. Experimental setup

The dynamics of the flux jumps in melt-grown HTS were studied in liquid nitrogen (T = 78 K). We measured the evolution of the induced current and the magnetic field during the pulsed field magnetization. After the magnetization the average critical current density was determined. Four annuli with outer diameter $2a_1 = 36$ mm, inner diameter $2a_0 = 17$ mm and thickness $2b \approx 11$ mm were studied. The annuli were cut from melt - grown Y-Ba-Cu-O discs which were prepared as described in ref. [3]. Critical temperatures of annuli were approximately identical; the critical current densities differed by 10-20 %. The trapped fields were $B_{tr} \approx 0.4$ T at $T = 78$ K. The magnetizing field was generated by copper coils located outside annuli. The current source allowed generating pulses with a peak field up to 3.5 T and a duration up to 40 ms. The magnetic field was measured by a Hall probe which was connected in parallel to the digital oscilloscope and to the DC micro-voltmeter. This setting allowed us to measure the field evolution during the pulse and field value after the magnetization with high accuracy. For the measurement of circular shielding current a Rogowski belt was used.

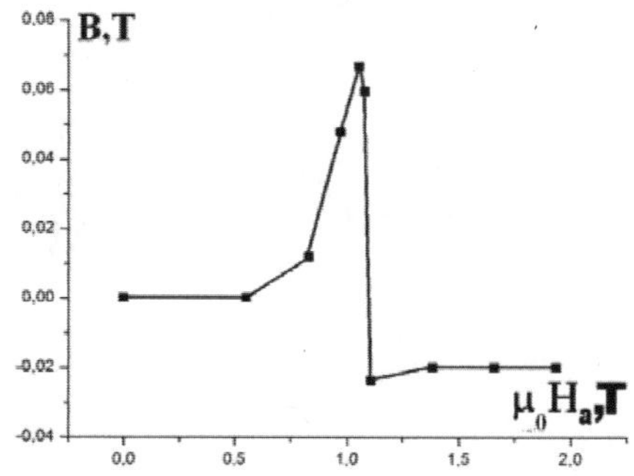

Figure 1. The trapped field in center of annuli as a function of external field peak.

3. Results and Discussion

In figure 1 the dependence of trapped field in the center of annuli versus the external field peak is shown. After each magnetization the annuli were heated above T_c. At low pulse peak ($\mu_0 H_a < 0.5$ T) the trapped field is absent ($B_{tr} \approx 0$). At these conditions circular currents screen the external field. With increasing the field peak B_{tr} appears and rises quickly. At some critical field peak $\mu_0 H_a \approx 1.0$ T the trapped field has a maximum and then falls sharply and changes the direction (sign). The sharp drop of B_{tr} and reversing its direction is a signature of a flux jump. Obviously, the field drop happens as a result of partial or full destruction of the shielding current circulating around the hole.

In figure 2 the evolutions of the shielding current and magnetic fields during flux jump are presented. The right axis relates to the external field of $\mu_0 H_a \approx 2$ T (dots) and the field in the center of the annuli (solid curve). The left axis relates to the shielding current evolution curve $I(t)$. Up to $t = t_c$ the current rises and then falls sharply by a factor of 10. On the increasing part of current, the resulting field is equal to 5-7 % of the external field. The sharp drop in the shielding current opens free way for the magnetic flux to penetrate into the hole of annuli. This is the flux jump. The Hall probe shows that starting with this moment of time the field in the center becomes nearly the same as the external field. The current drop is described by the exponential function $I = I_C exp(-t/\tau_f)$, where $\tau_f = L/R_{flow}$, (L – the inductance, R_{flow} – the flux flow resistance). The I_C characterizes the critical current. The characteristic time of current drop is $\tau_f = 0.36$ ms. This time remains practically the same for the magnetic field up to 5 T and for pulse duration down to 10 ms. In order to determine the R_{flow}, the current damping time was measured in pure aluminum annuli having identical sizes. It turned out that for Al (at $T = 78$ K) the decay time of the current is order of magnitude higher and is $\tau_{Al} = 3.6$ ms. Since the resistivity of Al at $T = 78$ K is $\rho_{Al} = 0.2 - 0.3$ μΩ·cm [5], the resistance of the Al annuli is $R_{al}=2\pi \cdot \rho_{A\lambda}(b \cdot ln(a_1/a_0))^{-1} = 2.5$ μΩ. Accordingly, the flux flow resistance in the superconductor is $R_{flow} = R_{Al} \cdot \tau_{Al} / \tau_f = 25$ μΩ.

During the decrease of the external field the shielding current remains small and the field in the hole is nearly equal to the external field. The low current value after the flux jump and the low rate of current variation is explained by the high temperature rise, which is supported by a high heat capacity and low heat conductivity of HTS. The thermal relaxation time in our samples is $\tau_{th} \approx 1$ s [6], explaining why the flux dynamics remain high and the difference between the external and the internal field is practically absent during the pulse.

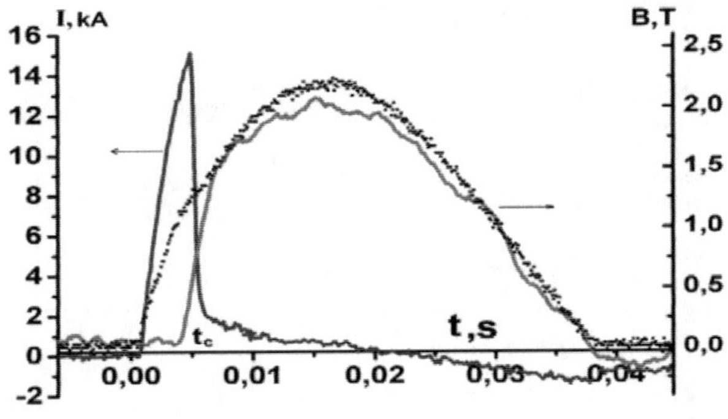

Figure 2. The evolution of current and field. Left axis – the current in annuli – I(t). Right axis – external field (dots) and field in the opening (solid curve).

The destruction of the critical current and the flux jump, as it is well-known, is caused by the overheating. In the experiment (figure 2) the critical current falls by 10 times. By using the linear temperature approximation of current I(T), the heating on $\Delta T_j \approx 10$ K is required. The heat dissipated during the jump is $q_j = \int I^2 \cdot R_{flow}\, dt \approx 1.28$ J. If this heat were dissipated over the whole volume of the annulus, the average temperature rise $<\Delta T> = q_j/\, (C_\Omega \cdot \cdot\, \Omega)$ (where C_Ω is the volumetric heat capacity [7]) would be less than 0.2 K. But this low heating does not correspond to the observed current drop during the flux jump.

To find out the distribution of the current in the superconducting annuli the radial field distribution $B_z(r)$ was measured in the central plane z = 0. For the measurements two identical annuli with 2 mm gap were used. A Hall probe was moved along the radial direction. In this geometry the axis component of the field was measured. From these data we can estimate the current density using the expression $J = \mu_0^{-1} dB_z/dr$. In figure 3 the flux density distribution $B_z(r)$ in the gap between annuli is shown for different magnetization conditions. The curve (1) corresponds to the single pulse magnetization with field peak of $\mu_0 H_a$ = 0.9 T. The curve (2) corresponds to the multi-pulse magnetization (10 pulses). The curve (3) shows the field distribution after the flux jump at $\mu_0 H_a = 2.0$ T.

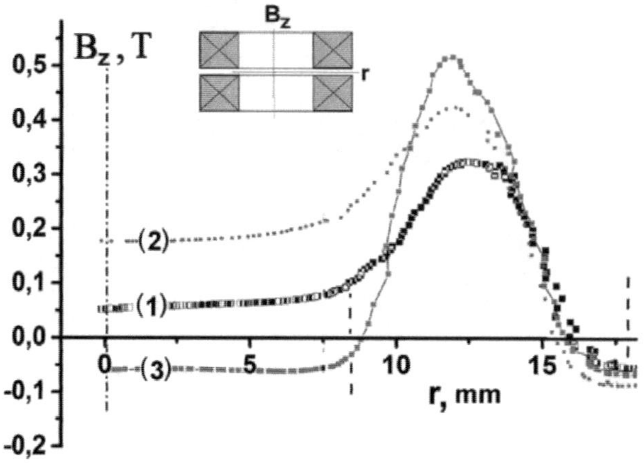

Figure 3. The radial field distribution $B_z(r)$ at the different conditions: (1) – after the single pulse of $\mu_0 H_a = 0.9$ T; (2) – after 10 pulses with peak field $\mu_0 H_a = 0.9$ T; (3) after the flux jump $\mu_0 H_a = 2.0$ T. The dotted line corresponds to the inner and outer annuli radii.

Currents in the region of $dB_z/dr < 0$ produce the field in the direction of the magnetizing field while the internal currents ($dB_z/dr > 0$) produce the field with the opposite direction. As shown in figure 2 when the current falls magnetic flux penetrates into the hole. When the external field is reduced, the flux trapping can be observed if I_c is high. Otherwise (if I_c is small) the flux escapes the hole and low field will be trapped. In the experiment the trapped field is low and has a direction opposite to that of the applied field. On the other hand from the figure 3 it follows that the critical current density $J_c = \mu_0^{-1}$ dB_z/dr is high enough 15 kA/cm^2. In fact it is order of magnitude higher than what remains after the flux jump. The high critical current density corresponds to the relatively low (< 0.5 K) heating as it is observed in the experiments [6,8]. Low critical current after the field jump corresponds to high heating, no less than $\Delta T_j \approx 10$ K. It is possible only in the case if the heat is dissipated not in all volume but only in the region of low pinning force. The volume of region occupied by the moving fluxoid at the flux jump can be estimated from the relation

$$\Omega_{flow} = q_j / (C_\Omega \cdot \Delta T_j) \qquad (1)$$

From the symmetrical considerations this region is reasonable to imagine as the sector of annulus. Thus we can obtain that sector angle is of $\varphi = 2\pi \cdot \langle \Delta T \rangle / \Delta T \approx 7$ deg.

The local heat dissipation in the narrow sector during the flux jump is explained by the azimuthal inhomogeneity of the critical current. In the melt-grown Y-Ba-Cu-O the critical currents in the (ab) plane have angle anisotropy which manifests itself in the anisotropy of trapped magnetic field in discs [8]. This inhomogeneity of the critical current is related to the technology of the HTS growth. Even if the anisotropy is low (10–15 %) due to the high exponent of HTS current-voltage curve ($E \sim J^N$, here $N > 25$ [9]), the heating temperature in this so-called "soft flux" region would exceed the heating temperature of region with higher J_c. Thus, the most of energy will be dissipated in the narrow sector of annuli.

Knowing the volume of soft flux region we can estimate the dynamical resistance of the flux motion. As the resistive region occupies the narrow sector of the annuli

$$\rho_{flow} = (2\pi/\varphi) \cdot \rho_{Al} \cdot (\tau_{Al} / \tau_f) \approx 100 \ \mu\Omega\cdot\text{cm} \qquad (2)$$

This value is approximately two times lower than the resistivity in ab plane at the normal state [7].

Due to the short pulse duration compared to the thermal relaxation time the region of flux motion remains at the high temperature after the flux jump. The rest of the annulus temperature is lower and J_c maintains high. Falling field magnetizes the body of annuli as in case the annulus has a slit. When the external field is switched off, the residual magnetization exhibits a maximum in the body of the annulus and low negative value near its internal and external edges. The symmetrical configuration of magnetic field shown in figure 3 creates the negative value in the center of the annulus when the radius of the annulus is comparable to its thickness. Similar distributions of $B_z(r)$ were observed in annuli with slit after the field cooling process and field removal [11].

4. Conclusion

During the pulsed field magnetization of melt-grown HTS annuli, the circular currents are destructed and a flux jump takes place. Because of the slow thermal processes vortices retain high mobility and the magnetic flux is not trapped in the hole of the annulus. The low value of the induced current and high value of critical current density in the body of the annulus indicates that: (1) during the flux jump the magnetic field gets into the hole through a "soft flux" region (i.e. with low J_c) of the annulus, (2) the soft flux region is the narrow annulus sector where the pulse heating is at least one order of magnitude higher than average heating.

Acknowledgments.
One of authors V.S.K. appreciate for the financial support from the Foundation of Assistance Small Entrepreneurship in Science (FASIE). Contract number 1309ГУ1/2014.

References

[1] Minz R, Rachmanov L 1977 *Uspechi Fizicheskikh Nauk* (Rus) **121** 499
[2] Gao L, Xue Y Y, Meng R L and Chu C W 1994 *Appl. Phys. Lett.* **64** 520
[3] Nizhelskiy N A, Poluschenko O L, Matveev V A 2007 *Supercond. Sci. Technol.* **20** 81
[4] Montgomery D B "Solenoid Magnet Design", Wiley-Interscience, 1969
[5] Iwasa Y "Case Studies in Superconductor Magnets: Design and Operational Issues",US Springer, 2009
[6] Korotkov V S, Krasnoperov E P and Kartamyshev A A 2015 *J. Supercond. Nov. Magn.* **28** 2815
[7] Tsuchimoto M and Morikawa K 1999 *IEEE Trans. Appl. Supercond.* **9** 66
[8] Yanagi Y, Itoh Y, Yoshikawa M, Oka T et al 2005 *Supercond. Sci. Technol.* **18** 839
[9] Douine B, Bonnard C-H and Sirois F 2015 *IEEE Trans. Appl. Supercond.* **25** 8001008
[10] Ainslie M D, Fujishiro H, Ujiie T et al. 2014 *Supercond. Sci. Technol.* **27** 065008
[11] Zheng H, Claus H, Chen L et. al. 2001 *Physica C* **17** 350

Critical current density and trapped field in HTS with asymmetric magnetization loops

D Gokhfeld

Kirensky Institute of Physics SB RAS, Akademgorodok 50/38, Krasnoyarsk 66036, Russian Federation

E-mail: gokhfeld@iph.krasn.ru

Abstract. Applications of the extended critical state model are considered. The trapped magnetic field, the penetration field and the field dependence of the critical current density are analysed. The critical current density and the trapped field in superconducting grains depend on the grain size. Asymmetry of the hysteresis curves relative to the $M = 0$ axis is related to the scale of the current circulation.

1. Introduction

The critical state model was modified more than once to describe different magnetization loops of type II superconductors [1-3]. The extended critical state model (ECSM) [4-7] is recent modification of the critical state model describing variety of magnetization loops. ECSM suggests that the asymmetry of the magnetization loop relative to the $M = 0$ axis is due to the equilibrium magnetization of the surface layer. The total magnetization of a sample is the sum of the equilibrium magnetization of the surface layer and the nonequilibrium magnetization of the inner volume. In the surface layer, the vortices are not pinned due to the interaction with the screening currents and the surface [8, 9].

ECSM combined with the computation of the field distribution into the sample [5-7] gives new facilities for analysis of magnetic characteristics of bulk superconductors. This model is applicable to describe magnetization loops of different superconducting materials [5, 10-16]. Estimations with using ECSM are useful for constructing of promising application devises (trapped field magnets, magnetic shielding, levitation etc.).

In this work some application aspects of ECSM are considered. The trapped field in superconducting samples is described in subsection 3.1. Parameters estimated from fitting of experimental magnetization loops result in phenomenological field dependence of the critical current density (subsection 3.2) and formula for the full penetration field (subsection 3.3). Determining of the screening current scale is discussed in subsection 3.4.

2. Extended critical state model

A cylindrical sample with the length significantly larger than the size of the base is under consideration. The demagnetization factor of such sample can be taken as zero. A thin plate oriented along the magnetic field can be considered in similar manner [7]. The magnetization M of a type II superconductor is defined as the diamagnetic response $-H$ plus the averaged magnetic field inside the sample, here H is the external magnetic field. For an infinitely long cylindrical sample with radius R, which is coaxial with the external magnetic field, the magnetization is determined by the following expression

$$M(H) = -H + \frac{2}{\mu_0 R^2}\int_0^R rB(r)dr\,, \tag{1}$$

where r is the distance from the cylinder axis, B is the local magnetic field in the sample, μ_0 is the magnetic constant. The distribution of the magnetic field $B(r)$ (the flux density) inside the sample is calculated from Ampere's circuital law $\mathrm{d}B/\mathrm{d}r = \mu_0 j_c(B)$ following approach [17], here j_c is the critical current density.

The dependence $M(H)$ is computed with the calculated distributions $B(r)$ for all H. It takes into account that the field distribution in the surface layer does not depend on the magnetization prehistory. The surface layer with the equilibrium magnetization has the depth l_s about the depth of magnetic field penetration λ [8, 9]. The asymmetry of the magnetization loops is determined by the fraction l_s/R.

The main steps of the parameterization and the calculation of the magnetization loop of a superconductor in ECSM are enumerated in work [7]. Parameters are chosen from the best fit of the calculated dependences to experimental magnetization loops. The calculated $M(H)$ curves for different values of l_s/R are presented on figure 1. To compute these loops the irreversibility field H_{irr} is established to be much smaller than the critical field H_{c2}. Such case is relevant for high temperature superconductors (HTS) [7]. As the value of l_s/R growths from 0 to 1 the asymmetry of the loops increases. The fully reversible curve for $l_s/R = 1$ (maximal asymmetry) coincides with the virgin magnetization branches of all curves. The critical current density dependencies on the magnetic field $j_c(H)$ extracted from the $M(H)$ loops with using ECSM and the Bean model are presented on figure 2 (see subsection 3.2).

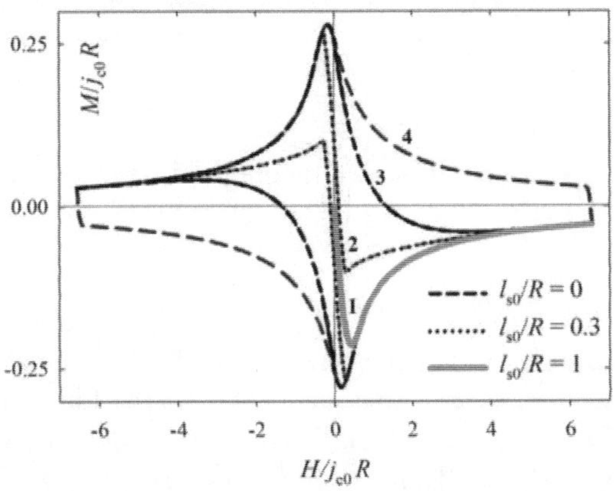

Figure 1. Magnetization loops $M(H)$ computed by ECSM. The curves with $H_{irr} = 10H_p$: reversible curve $l_{s0}/R = 1$ (1), $l_{s0}/R = 0.3$ (2), $l_{s0}/R = 0$ (3); the near symmetric hysteresis with $l_{s0}/R = 0$, $H_{irr} = 200H_p$ (4).

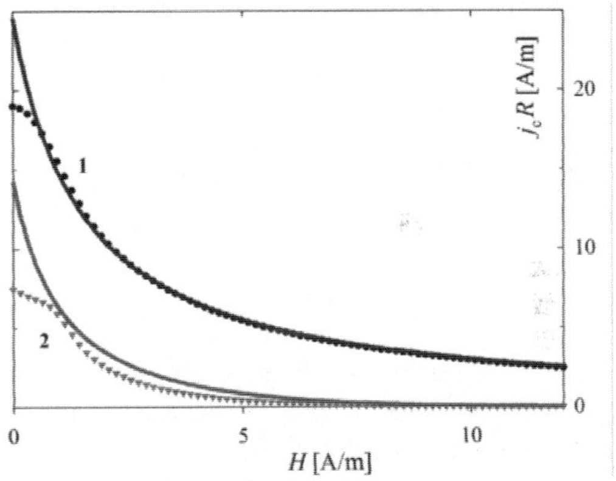

Figure 2. Critical current density $j_c(H)$ obtained from $M(H)$ loops. Curves 1 were obtained from the hysteresis 4 shown in figure 1 and curves 2 from the loop 2 using the Bean model (points) and ECSM (lines) with $n = 2$.

3. Applications

ECSM is useful for simple estimations of various parameters of superconducting samples. The model considered is applicable mainly to samples in the form of a long cylinder or a plate, for which the demagnetization factor is zero. But the penetration of the magnetic flux into samples of other shapes at high magnetic fields occurs in much the same manner [17] and the model can also be used for approximate estimations. Polycrystalline and heterogeneous superconductors are also subjects of

ECSM (see subsection 3.4). The fishtail effect on magnetization loops or the peak effect [18] can also be treated by ECSM [6, 14-16]. For this purpose, it is necessary to determine the function describing the peak in the dependence $j_c(H)$ and the corresponding dip in the dependence $l_s(H)$ [6].

3.1. Trapped field

The computed distributions $B(r)$ gives the magnetic field in the middle of the sample and the trapped field distribution. The dependence of the magnetic field in the sample middle on the external magnetic field is presented on figure 3. The maximal value of the trapped field is equal to the full penetration field H_p. Increase of the ratio l_s/R leads to a decrease of the trapped field in the sample middle.

The spatial distribution of the field trapped in the sample after the field cycle is drawn on figure 4. The trapped field follows a conic profile. Increase of l_s/R sinks this cone (not displayed).

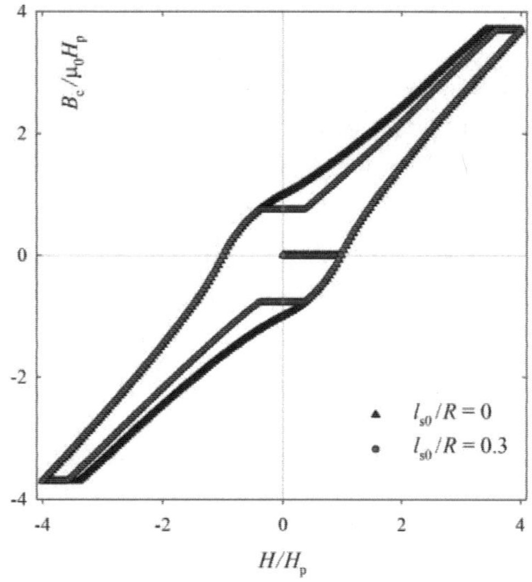

Figure 3. Magnetic field B_c in the sample middle.

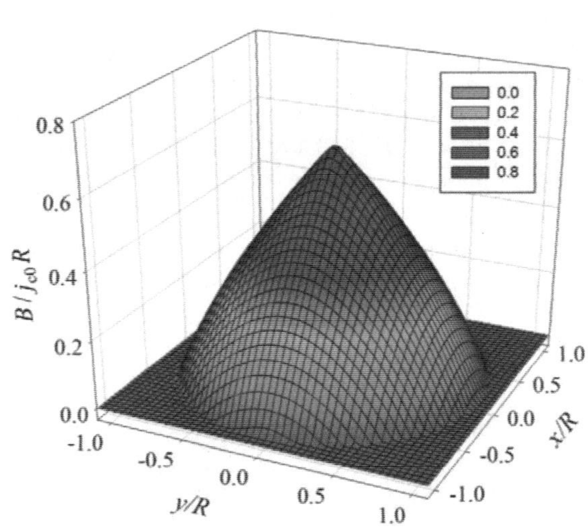

Figure 4. Calculations of magnetic field B trapped in the sample.

3.2. Critical current density

The Bean model is commonly used to determine the critical current density j_c and its field dependence from magnetic measurements. However it can give incorrect results in some magnetic field ranges, e.g. in the vicinity of $H = 0$. An alternative method is to fit different analytical expressions for $j_c(H)$. The relevance of the dependence $j_c(H)$ is judged from agreement between the computed and experimental magnetization loops. Simple dependencies $j_c(H) = \text{const}$, $j_c(H) \sim 1/H$ and $j_c(H) \sim \exp(-H)$ do not lead to agreement between the calculated magnetization loop and the experimental data in strong and weak fields simultaneously. Based on many performed fittings, the dependence $j_c(H)$ was chosen, which is proportional to $1/H$ in weak fields and decreases exponentially in strong fields,

$$j_c(H) = j_{c0}\left[|H|/H_1 + \exp\left(|H|/H_2\right)\right]^{-1}, \tag{2}$$

where j_{c0} is the value of $j_c(H)$ for $H = 0$, H_1 and H_2 are the parameters specifying the scales of field decreasing. Given $H_2 \sim 0.1H_{c2}$, the calculated values of j_c tend to zero for $H \geq H_{c2}$.

For near symmetric magnetization loops, the dependence $j_c(H)$ expressed by function (2) coincides over the entire field range, except for weak fields H, with the Bean model curve (see curves 1 on figure 2). In this case the contribution from the equilibrium magnetization of the surface layer is negligible because $l_s << R$. If values of l_s and R are comparable then the surface layer should be accounted. The surface layer of the superconducting sample with the thickness l_s is not involved in supercurrent transport. The surface layer depth depends on the external magnetic field and the temperature. The simple form is used for the dependence $l_s(H)$ [7]: $l_s(H) = l_{s0} + (R - l_{s0}) H/H_{irr}$, where l_{s0} is the value of l_s at $H = 0$, H_{irr} is the irreversibility field. When the critical current density is estimated, the area of the surface layer should be omitted such that the averaged critical current density $j_c(H)$ becomes dependent on R. The following dependence $j_c(H)$ is suggested

$$j_c(H) = j_{c0}\frac{\left[(1 - l_{s0}/R)(1 - H/H_{irr})\right]^n}{\left[|H|/H_1 + \exp(|H|/H_2)\right]}, \tag{3}$$

here n is the form-dependent power [7]: $n = 2$ for a cylindrical sample and $n = 1$ for a thin plate oriented along the field. With accounting of the screening current one should use $n = 3$ for a cylindrical sample and $n = 2$ for a thin plate. Figure 2 demonstrates that account of l_s for $n = 2$ is modified the dependence $j_c(H)$. The ECSM curves for $n = 3$ coincide with the curves obtained by the Bean model.

3.3. Full penetration field

In ECSM the full penetration field H_p is calculated numerically and the resulted values of H_p are some smaller than the Bean model values of $H_p = j_cR$. Formula (4) gives approximate values of H_p.

$$H_p = \frac{2H_1H_2}{H_1 + H_2}\ln\left[1 + \frac{j_{c0}R(H_1 + H_2)}{2H_1H_2}\right]. \tag{4}$$

This analytic expression is obtained from formula (2) and fitting numerically computed values of H_p.

3.4. Scale of screening current

In formulae 3 and 4 the size R determines the circulation region of the screening current. For granular superconductors, there is an ambiguity in choosing the scale of R [12, 19]: It can be chosen as the radius of the sample or as the average effective radius of the grains. The screening current can also circulate in clusters consisting of several grains. In order to determine the critical current density from magnetic measurements, it is necessary to establish the scale of the circulation of the screening current, because the magnetization hysteresis width is characterized by the product j_cR. But the circulation radius of the screening current also defines the hysteresis asymmetry by means of the relation l_s/R. So given $l_{s0}(T) \approx \lambda(T)$, one can find the circulation scale of the screening current from ECSM fitting of an asymmetric magnetization hysteresis [7]. This procedure is useful to accurate estimating of the intragrain critical current density. For heterogeneous samples the porosity must be also accounted for estimating of j_c [5, 10]. For the granular superconductors analyzed earlier [5, 10, 12-14, 16], the effective radius satisfying the fitting is about the averaged grain size in the *ab* plane determined from scanning electron microscopy images.

4. Conclusion

Magnetization loops of type II superconductors are computed and parameterized by the extended critical state model. The magnetic flux distributions in the sample and the trapped field have been calculated. Asymmetry of magnetization hysteresis relative to the $M = 0$ axis originates due to the equilibrium magnetization of the surface layer. So asymmetry of the magnetization loop depends on the ratio of the surface layer depth to the radius of the screening current circulation which can be the sample size or the grain size. In the ranges of temperature and magnetic field where the depth of magnetic field penetration λ is comparable with the radius of the screening current circulation, the critical current density and the trapped field depend on l_s/R also. It is important for polycrystalline superconductors and for large grain superconductors at high magnetic fields and temperatures.

References

[1] Chen D X and Goldfarb R B 1989 *J. Appl. Phys.* **66** 2489
[2] Xu M, Shi D and Fox R F 1990 *Phys. Rev.* B **42** 10773
[3] Senoussi S, Aguillon C and Manuel P 1991 *Physica* C **175** 202; Senoussi S and Aguillon C 1990 *Proc. ICMAS-90* (*Grenoble*) (Ed. IITT) p 159
[4] Chen D X, Cross R W and Sanchez A 1992 *Cryogenics* **33** 695
[5] Gokhfeld D M, Balaev D A, Petrov M I, Popkov S I, Shaykhutdinov K A and Valkov V V 2011 *J. Appl. Phys.* **109** 033904
[6] Gokhfeld D M 2013 *J. Supercond. Novel Magn.* **26** 281
[7] Gokhfeld D M 2014 *Phys. Solid State* **56** 2380
[8] Burlachkov L 1993 *Phys. Rev.* B **47** 8056
[9] Blinov E V, Laiho R, Lahderanta E, Stepanov Yu P, Traito K B and Vlasenko L S 1994 *J. Exp. Theor. Phys.* **79** 433
[10] Gokhfeld D M, Balaev D A, Popkov S I, Shaykhutdinov K A, Petrov M I 2006 *Physica* C **434** 135
[11] Altin S and Gokhfeld D M 2011 *Physica* C **471** 217
[12] Altin E, Gokhfeld D M, Komogortsev S V, Altin S and Yakinci M E 2013 *J. Mater. Sci.: Mater. Electron.* **24** 1341
[13] Yakinci Z D, Gokhfeld D M, Altin E, Kurt F, Altin S, Demirel S, Aksan M A, Yakinci M E 2013 *J. Mater. Sci.: Mater. Electron.* **24** 4790
[14] Altin E, Gokhfeld D M, Kurt F and Yakinci M E 2013 *J. Mater. Sci.: Mater. Electron.* **24** 5075
[15] Balaev D A, Gokhfeld D M, Popkov S I, Shaikhutdinov K A, Klinkova L A, Zherikhina L N and Tsvokhrebov A M 2014 *J. Exp. Theor. Phys.* **118** 104
[16] Altin E, Gokhfeld D M, Demirel S, Oz E, Kurt F, Altin S, Yakinci M E 2014 *J. Mater. Sci.: Mater. Electron.* **25** 1466
[17] Forsthuber M and Hilscher G 1992 *Phys. Rev.* B **45** 7996
[18] Johansen T H, Koblischka M R, Bratsberg H and Hetland P O 1997 *Phys. Rev.* B **56** 11273
[19] Angadi M A, Caplin A D, Laverty J R and Shen Z X 1991 *Physica* C **177** 479

Waveform control pulse magnetization for HTS bulk magnet

Tetsuya Ida[1], Koji Shigeuchi[2], Sayo Okuda[3], Masahiro Watasaki[4] and Mitsuru Izumi[5]

[1] National Institute of Technology, Hiroshima College, Hiroshima 725-0231, Japan
[2] Chiba University, Chiba 263-8522, Japan
[3] Toyohashi University of Technology, Aichi 441-8580, Japan
[4] Sumitomo Chemical Co., Ltd., Ehime 792-8521, Japan
[5] Tokyo University of Marine Science and Technology, Tokyo 135-8533, Japan

E-mail: ida@hiroshima-cmt.ac.jp

Abstract. For the past 10 years, we have studied high-temperature superconducting (HTS) bulk magnets for use in electromagnetic rotating machines. If the magnetic field effectively magnetizes the HTS bulk, then the size of the motor and generator can be reduced without a reduction in output. We showed that the melt-textured Gd-Ba-Cu-O HTS bulk effectively traps a high magnetic field using waveform control pulse magnetization (WCPM). WCPM makes it possible to generate any pulsed magnetic field waveform by appropriately changing the duty ratio of the pulse width modulation. By chopping so that the pulsed magnetic field has a period of about 1ms, the WCPM technology enables active control of the rise time and suppresses magnetic flux motion that decreases magnetization efficiency. This method is also useful for any HTS bulk magnet, and the high magnetic flux density is trapped in the HTS bulk by a single pulse magnetic field. We developed a magnetizer that has a feedback system from the penetrated magnetic flux density to realize WCPM. In this research, using only a single pulse magnetic field of WCPM method at 77K, an HTS bulk with a 45mm diameter and 19mm thickness trapped a maximum magnetic field of 1.63T, which is more than 90% of the trapped magnetic flux density by FC magnetization. This result suggests that the pulse magnetizing method can replace the conventional field-cooled method and promote the practical use of HTS magnets for electromagnetic power applications.

1. Introduction

High-efficiency energy consumption is needed for industrial electromagnetic power applications involving rotating machinery. However, the low residual magnetic flux density of magnets complicates the realization of such high efficiencies. A high-temperature superconductor (HTS) bulk with a strong pinning effect can trap much higher magnetic fields than is possible with conventional permanent magnets [1, 2]. Accordingly, the HTS bulk can contribute to the performance enhancement of next-generation electromagnetic rotating machines.

For practical use, a pulsed field magnetizes the HTS bulk installed in an electromagnetic rotating machine using a small metal coil. However, the magnetic flux density that the HTS bulk traps by pulse field magnetization (PFM) is lower than that trapped by field-cooled (FC) magnetization [3]. By applying a strong pulsed field, the trapped magnetic flux density distribution is distorted asymmetrically and it decreases due to the large amount of heat

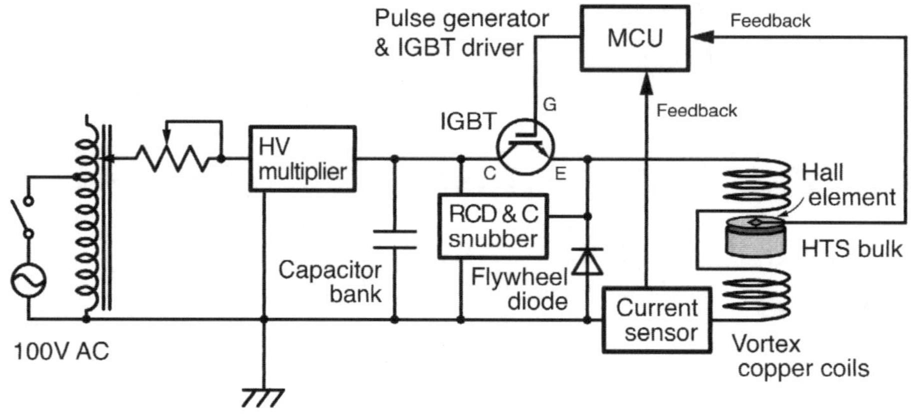

Figure 1. Electrical block diagram of pulse magnetizer for WCPM method.

generation originating in the dynamic transient flux flow upon pulsed field magnetization [3–5]. The suppression of the trapped magnetic field consisting of the transient heat from the magnetic flux motion must be improved to achieve sufficient magnetization for applications [6]. It is important that HTS motors suppress the magnetic flux motion caused by PFM in order to obtain a high magnetic field, because the trapped magnetic field of the superconductor depends on the magnetic flux pinning force [7]. Therefore, we are interested in the motion of magnetic flux in melt-processed RE-Ba-Cu-O (RE indicates a rare earth element) bulk superconductors for use in advanced industrial power applications. In general, the pulsed field depends on an exponential function that is generated by a passive capacitor discharge. By the multi-peak PFM method, several exponential pulsed fields enhance the trapped magnetic field in the HTS bulk [8,9]. Because the method of magnetization takes a long time to magnetize the HTS bulk, we need a simple and practical magnetization technology to start the HTS motor.

A pulsed field generated by a passive element is not suitable for magnetizing the HTS bulk. Accordingly, we aim to produce a pulsed field that is adapted to the HTS bulk by active control. For the past decade, we have studied a new method of HTS bulk magnetization that uses the pulse magnetic field of an arbitrary waveform that does not depend on the conventional LCR transient response [10–12]. The waveform control pulse magnetization (WCPM) method distorts the pulsed field waveform to suppress discharge using chopping [13]. The single pulse field could suppress the magnetic flux motion and increase the trapped magnetic fields. The melt-textured Gd-Ba-Cu-O bulk was magnetized to an excellent cone-shaped magnetic field distribution. To increase the trapped magnetic flux density by the applied pulse magnetic field, we developed a magnetizer with a feedback circuit that controls the magnetic flux density in the bulk. In our previous work, we used this control to maintain a constant peak current when the penetration magnetic flux density in the HTS bulk exceeded a threshold close to the critical magnetic field [12]. The sequential style feedback control based on this electrical current could not eliminate the influence of the magnetic flux that already penetrated the HTS bulk, so we could not suppress the flux jump. In this study, we applied a magnetizing pulse based on the penetration magnetic flux to the HTS bulk in order to obtain a high magnetic field density, and aimed to suppress the flux jump.

2. Pulse magnetization

When the switch included in the pulse magnetizer is closed, a discharging current flows through the magnetizing coil from the capacitor bank. The discharging current will gently decrease due to the effect of electromagnetic induction if the switch opens. Thus, we can control the pulsed

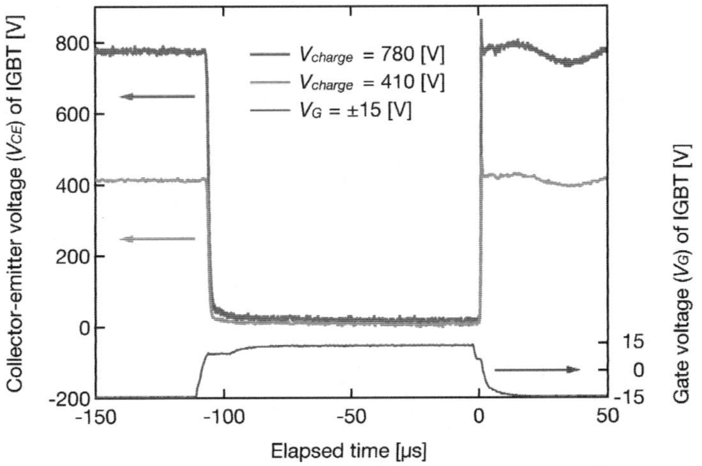

Figure 2. IGBT switching characteristics.

Figure 3. Pulse magnetizer with magnetic field feedback circuit for WCPM method.

current by chopping as we repeat the switching operations over a short time. The switching occurs over a short period by applying a high self-induced voltage to the insulated gate bipolar transistor (IGBT) of the chopper. We choose an IGBT that has maximum ratings of DC 1700V/2400A so that the high voltage does not break down the IGBT. Furthermore, we made a snubber circuit for the pulse magnetizer. Accordingly, we switched the chopper with a period of 1ms, the same as in previous work. In this study, we developed pulse magnetizing equipment that controls a pulse magnetic field by feedback input of the penetration magnetic flux. Figure 1 shows an electrical block diagram of the pulse magnetizer developed for the WCPM method. This magnetizing equipment generates maximum DC 1kV from AC 100V. The three parallel IGBTs discharge electrical energy up to 25kJ charged to a capacitor bank of $50,000\mu$F in total. The reverse voltage to be generated between the collector and emitter of the IGBT at the time of chopping would break the IGBT. Therefore, we mounted a flywheel diode and a snubber circuit to protect the IGBT. Figure 2 shows the collector-to-emitter voltage by turning on the IGBT for 100μs following the rise/fall of the gate voltage. At the time of breaking, the switching of IGBT produced a surge voltage of 467V from the capacitor bank voltage of 410V at $t = 0$, and produced a surge of 864V from 780V the same way. The surge voltages are only 12% larger than the capacitor bank voltages. With various methods, our pulse magnetizer can control DC 5kA or 10kA of pulsed current in 1kV or less, without breaking.

The IGBT driver applies a gate voltage to the IGBT according to an input gate signal from a 16-bit micro control unit (MCU). The gate pulse with that duty ratio was changed into a controls magnetizing current. The input gate signal is determined as well as a sequential program by a feedback signal of magnetizing current and magnetic flux density. The 16-bit analog-digital converters (ADCs) measure each of the pulse magnetizing currents and the penetration flux density, and it sends them to the MCU.

The magnetizer for the WCPM method is shown in Fig. 3. A chopper circuit is placed at the bottom of the three levels of structures. The middle layer is the capacitor bank for charging the electric power, and the upper layer is the booster circuit and pulse controller with the MCU. This electronics device limits the charging current, controls the charging voltage, and controls the magnetizing current using commands from the controller automatically.

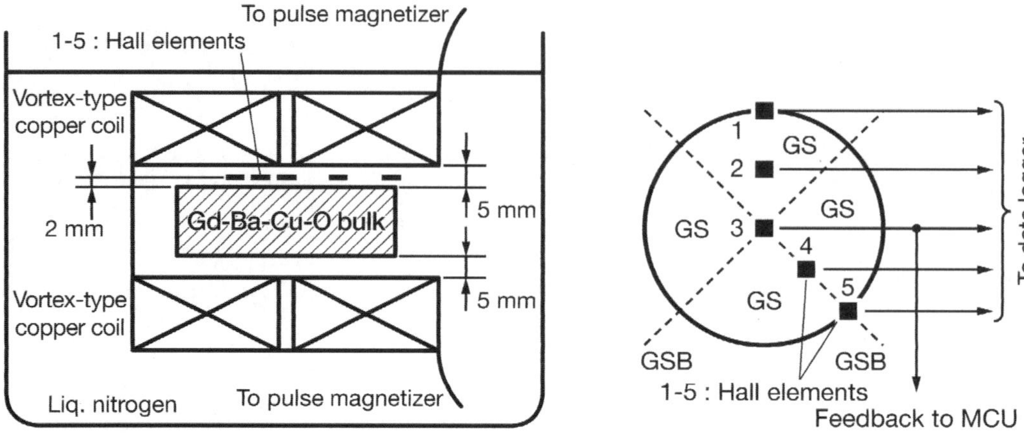

Figure 4. Bulk sample geometry for pulse magnetization and measurement.

3. Experiment and result

A highly c-axis oriented Gd-Ba-Cu-O bulk sample ($GdBa_2Cu_3O_{6.9}$ 70.9wt%, Gd_2BaCuO_5 19.2wt %, Ag 9.4wt%, Pt 0.5wt %, QMG, NSSMC) which was made by the melt growth process was used for measurements [11]. The sample was 45mm in diameter and 19mm in thickness. Using the static magnetic field in FC magnetization at 77K, we obtained a maximum trapped flux density of 1.7T with a conical distribution on the sample surface. In this study, the sample was immersed in liquid nitrogen for the pulse magnetization. The magnetic flux motion in the bulk depends on the crystal orientation. Thus, we put five Hall elements on the bulk center, growth sector, growth sector boundary and the outer edges. The details of the sample geometry are shown in Fig. 4. The Hall elements are THS118 (Toshiba) with a GaAs semiconductor type. In this study, we measured the magnetic flux density for feedback to the MCU by using the Hall element in the bulk center. The vortex-type copper coil, which has an inductance of 650μH, has a diameter of 84mm and a 20mm thickness [3].

We magnetized the HTS bulk sample by WCPM at 77K. The applied pulsed magnetic field generated from the excessive electric power of 12.3kJ was controlled by the feedback of the magnetic flux measured at the HTS bulk surface center in the magnetization, shown in Fig. 5. The chopping period of the applied pulse magnetic field was 1kHz. It is likely that a large number of impulses that are included in the magnetic flux density waveform are due to switching noise by the fluctuation of the high current. However, we consider that the impulses are provided as a result of the feedback control, as we mention it later. When the flux density penetrating the bulk center exceeds a preset threshold, the MCU reduces the duty ratio of the gate pulse sent to the IGBT. In contrast to proportional-integral control in Figs. 6 and 7 of Ref. [12], we tried proportional-derivative (PD) control to follow the steep variation of the magnetic flux. The MCU, which performed PD control by prior input of the proportional gain and the derivative gain, varied the duty ratio depending on the penetration magnetic flux density. In this experiment, we set the integral gain to 0. However, the waveform control signal was integrated every 1ms for chopping. With such a control, the pulse controller suppressed the electric current supplied to the vortex-type copper coils depending on the penetration magnetic flux density, until the pulse magnetization was finished. Because the electrical energy of the capacitor bank was discharged little by little, most of the pulsed magnetic field was generated for 2 seconds. A low pulsed magnetic field continuously was applied for about 4 seconds.

In a previous study, we showed that an instability such as a flux jump occurs around 1.7T of the critical magnetic field by WCPM, which limits the current with the magnetic flux density

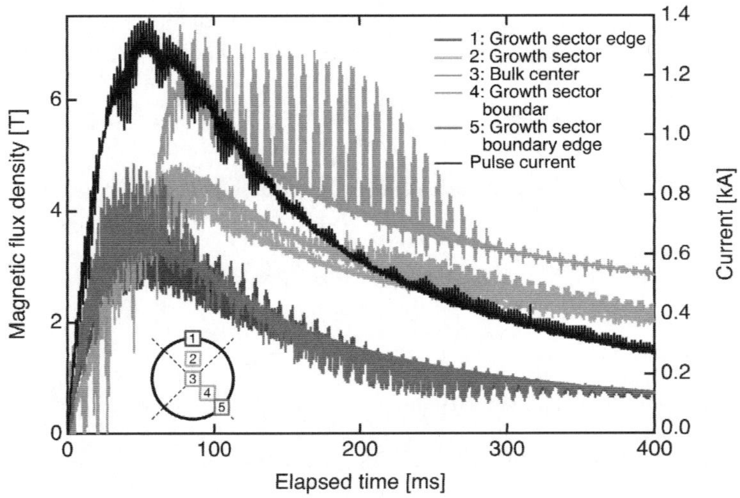

(a)

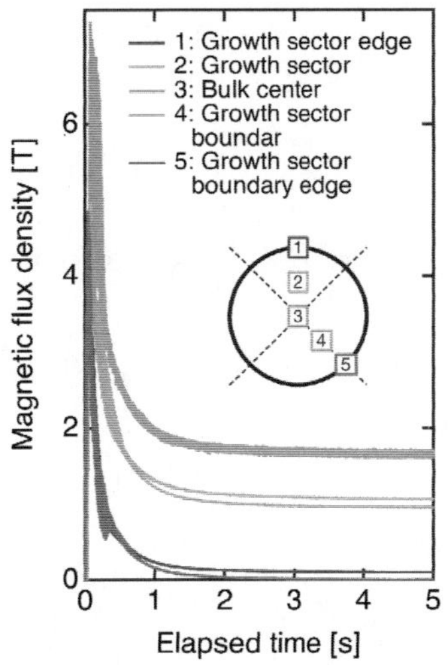

(b)

Figure 5. Magnetic flux density on the bulk (a) for short elapsed time from a start of the magnetization, and (b) after 5 seconds elapsing. Both insets show geometry of 1-5 Hall elements.

of the threshold [11, 12]. The magnetic flux motion seems to be strongly inhibited below this threshold, and the magnetic flux density in the bulk increased suddenly when the control did not work. In this study, we set the threshold to 1.8T, which was slightly higher than 1.7T used in Ref. [12]. However, the magnetic field greatly exceeds 1.8T penetrated to the bulk. The reason may be due to the slowness of the integral time of the ADC, which is 1ms on the feedback control. When the penetration magnetic flux density exceeds a critical state, a flux jump will occur. The slow control cannot completely suppress a sudden increase in the magnetic flux density caused by the flux jump. Many of the impulses included within the curves in Fig. 5(a) may be flux jumps that were not suppressed because of incomplete feedback waveform control. Nevertheless, the trapped magnetic field reached 1.63T at the surface center and had a tendency to be distributed into a conical shape. This experimental result for the single PFM realized more than 90% of the magnetic flux density for the critical magnetic field of 1.7T. We consider that the suppression of the incomplete flux jump had the effect of trapping the high magnetic field on the bulk. Otherwise, our HTS bulk might be magnetized by a single WCPM field for approximately 4 seconds, which was generated by an electric power of 12.3kJ. There is a possibility that such an applied magnetic field over a long time would increase the trapped flux density.

4. Summary

We made a pulse magnetizer to magnetize the HTS bulk by the WCPM method. The magnetizing equipment has the ability to measure the magnetic flux density that penetrated the bulk, and the discharge circuit without auto-destruction was possible by the surge voltage. The magnetic flux density at the HTS bulk center fed back to the MCU-controlled amplitude of the pulsed field applied to the Gd-Ba-Cu-O bulk. We magnetized the HTS bulk to 1.63T by a single pulsed field at 77K for the threshold of 1.8T. This magnetizing condition was not able to suppress the flux jump. However, this is beyond 90% of the trapped magnetic flux density by FC magnetization. The very effective magnetizing of the bulk leads to practical uses of HTS materials in electromagnetic power applications. In the future, we will clarify the condition for trapping a high magnetic flux density to various superconducting materials by studying the details of the magnetizing process together with a trial to suppress the flux jump. And a high magnetic field at a low temperature will be trapped using the single pulse magnetic field of WCPM.

Acknowledgments
This work was supported by JSPS KAKENHI Grant Numbers 23360394,15K05971 (2011-2013, 2015-), and Electric Technology Research Foundation of Chugoku (2010). We thank Dr. Motohiro Miki of the TUMSAT for technical support.

References
[1] Tomita M and Murakami M 2000 *Supercond. Sci. Technol.* **13** 722
[2] Tomita M and Murakami M 2003 *Nature* **421** 517
[3] Ida T, Matsuzaki H, Akita Y, Izumi M, Sugimoto H, Hondou Y, Kimura Y, Sakai N, Nariki S, Hirabayashi I, Miki M and Kitano M 2004 *Physica C* **412-414** 638
[4] Kono T, Tomita M, Murakami M and Takizawa T 2000 *Supercond. Sci. Technol.* **13** 794
[5] Yanagi Y, Itoh Y, Yoshikawa M, Oka T, Hosokawa T, Ishihara H, Ikuta H and Mizutani U 2000 *Adv. Supercond.* **12** 470
[6] Sander M, Sutter U, Adam M and Kläser M 2002 *Supercond. Sci. Technol.* **15** 748
[7] Matsuzaki H, Kimura Y, Ohtani I, Izumi M, Ida T, Akita Y, Sugimoto H, Miki M and Kitano M 2005 *IEEE Trans. Appl. Supercond.* **15** 2222
[8] Ikuta H, Yanagi Y, Yoshikawa M, Itoh Y, Oka T and Mizutani U 2001 *Physica C* **357-360** 837
[9] Fujishiro H, Kaneyama M, Tateiwa T and Oka T 2005 *Jpn. J. Appl. Phys.* **44** L1221
[10] Ida T, Matsuzaki H, Morita E, Sakashita H, Harada T, Ogata H, Kimura Y, Miki M, Kitano M and Izumi M 2006 *J. Phys.: Conf. Series.* **43** 539
[11] Ida T, Kimura Y, Sano T, Yamaguchi K, Izumi M and Miki M 2008 *J. Phys.: Conf. Series.* **97** 012292
[12] Ida T, Watasaki M, Kimura Y, Miki M and Izumi M 2010 *J. Phys.: Conf. Series.* **234** 032023
[13] Tokyo Univ. of Marine Sci. and Technol., National Institute of Technology, Hiroshima College, Kitano Seiki Co. Ltd., JP Patent 4793801 B, 2011-10-12

Growth conditions and properties of Y123/Y211 and Y123/Y_2O_3 bulks

M Radušovská, D Volochová, V Antal, S Piovarči and P Diko

Institute of Experimental Physics, SAS, Watsonova 47, 040 01 Kosice, Slovakia

E-mail: radusovska@saske.sk

Abstract. The single-grain YBCO samples were prepared by TSMG process. Two nominal compositions: 1 mol.Y123 + 0.25 mol.Y_2O_3 + 1.0 wt.% CeO_2 (Y123/Y_2O_3) or 70 wt.% Y-123 + 30 wt.% Y-211 (Y123/Y211) and 1.0 wt.% CeO_2 and two different types of substrates were tested. It is shown that the growth rate in both systems is influenced by segregation of CuO in to the nonsolidified part of the sample what causes stop of isothermal growth. Measured melt losses are lower for ZrO_2 substrate than for Y_2O_3/Yb_2O_3 substrate what influence final microstructure of prepared bulk superconductors.

1. Introduction
Currently, the number of scientific publications on REBCO bulk single-grain superconductors is focused on the details of preparation parameters to optimize technology of TSMG process [1-6]. The growth conditions are also intensively studied mainly in the systems with cheap CeO_2 addition for refinement of Y211 particles [7-13]. This paper compares Y123/Y211 and Y123/Y_2O_3 systems with the addition of 1wt.% CeO_2. It is shown that mainly the appearance of CuO due to reaction of 123 with Y_2O_3 and CeO_2 influences growth of single-grain samples. It is also shown that higher 211 concentration in the sample with high melt losses causes difficulties at oxygenation.

2. Experimental
YBCO bulk samples with two nominal compositions: 1 mol. Y123 + 0.25 mol. Y_2O_3 + 1.0 wt. % CeO_2 and 70 wt.% Y123 + 30 wt.% Y211 + 1.0 wt.% CeO_2 were prepared in air in a chamber furnace. $YBa_2Cu_3O_7$ (Solvay, particle size 30μm), Y_2O_3 or Y211 (Solvay, particle size 5μm) and CeO_2 powders were mixed in appropriate amounts in a mixer for 30 min and then intensively milled for 15 min in a friction mill. The homogenized milled mixture of powders was uniaxially pressed into cylindrical pellets 20 mm in diameter. The samples were treated in a chamber furnace according to the following steps: heating up at the rate of 100 °C/h to 940 °C/dwell 24h, heating up at the rate of 100 °C/h to the maximum temperature, T_{max} = 1040 °C (30 °C over peritectic temperature, T_p), dwell time for 1h, fast cooling to a temperature 12 °C higher than the temperature of isothermal grow, T_{is}, at the rate of 50 °C/h, slow cooling to the isothermal hold temperature, T_{is}, with cooling rate 1 °C/h and finally cooling to the room temperature with furnace. The macrostructure of the sample surfaces was studied by a stereomicroscope. Sample cuts in the *a/c*-plane along the cylinder axis were prepared by grinding and polishing for microscopic analyses. The microstructure details were analyzed in an optical microscope under normal and polarised light. DTA measurements of mixed nominal powder and pure $YBa_2Cu_3O_7$ powder in the artificial air were applied to determine characteristic transformation temperatures. The

samples were oxygenated in flowing oxygen gas at 400°C for 150h. For trapped field measurements, the samples were field cooled in liquid nitrogen (77K) in the presence of an applied external magnetic field of 1,4T.

3. Result and discussion

The single-grain Y123 samples with different compositions were successfully grown by Top-Seeded Melt-Growth (TSMG) process. Macroscopic images of the top surface of grown single-grains are presented in Figure 1.

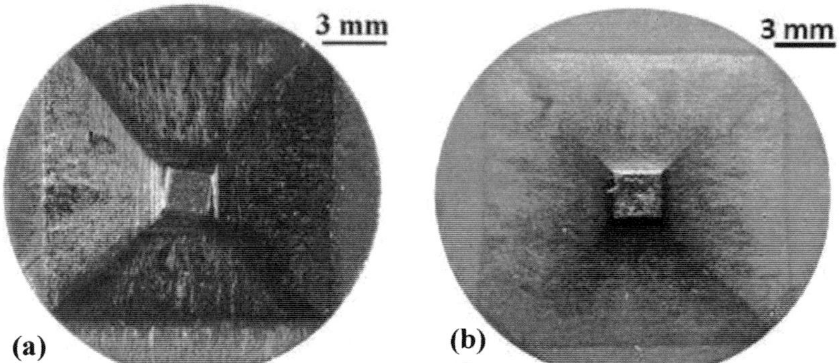

Figure 1. The top surface of the sample with 60 hours isothermal dwell at 996 °C (a) and the top surface of the sample with 80 hours isothermal dwell at 992 °C (b).

The used times of isothermal dwell were long enough for the single-grain to grow fully through the sample diameter if the growth rate 0.2 mm per hour, measured in the middle of sample radius, is considered [14]. The observed growth stop is the result of CuO appearance in the system due to the reaction of Y_2O_3 and CeO_2 with Y123 and consequent CuO segregation into the nonsolidified rest of melt [14-15]. The presence of CuO in the Y123/Y211 system with CeO_2 addition is also confirmed by lower peritectic temperature of this system measured by DTA (Figure 2(a)) and high CuO concentration in the solidified melt at the sample rim (Figure 2(b)).

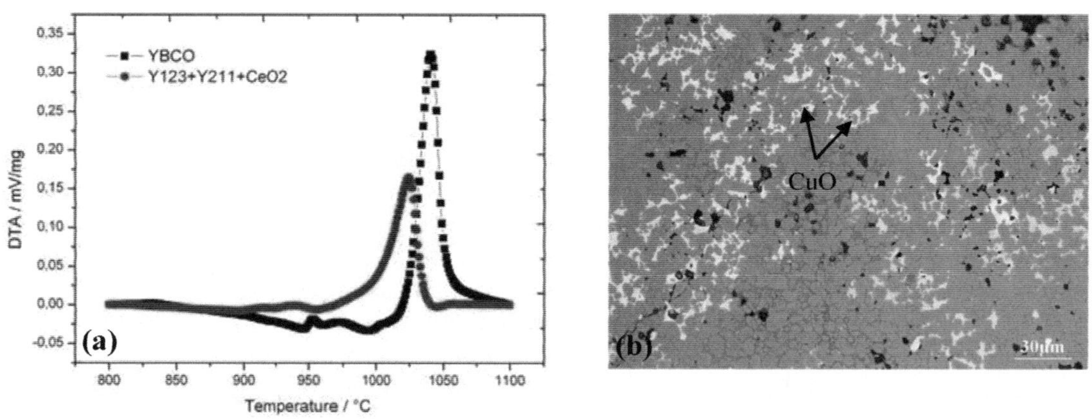

Figure 2. DTA records at heating with 10°C/min. in artificial air for pure $YBa_2Cu_3O_x$ powder and mixture of Y123/Y211 with addition 1wt.% CeO_2 (a) and CuO phase (bright) present in the Y123/Y211 system in the solidified melt at the sample rim (b).

After oxygenation of prepared bulk superconductors the measured trapped field profiles at 77K show significant differences (Figure 3). While the sample grown from the Y123/Y_2O_3 system exhibits beautiful symmetric cone (Figure 3(b)), the trapped field profile of the sample prepared from the Y123/Y211 system is highly irregular with much lower maximum trapped field (Figure 3(a)).

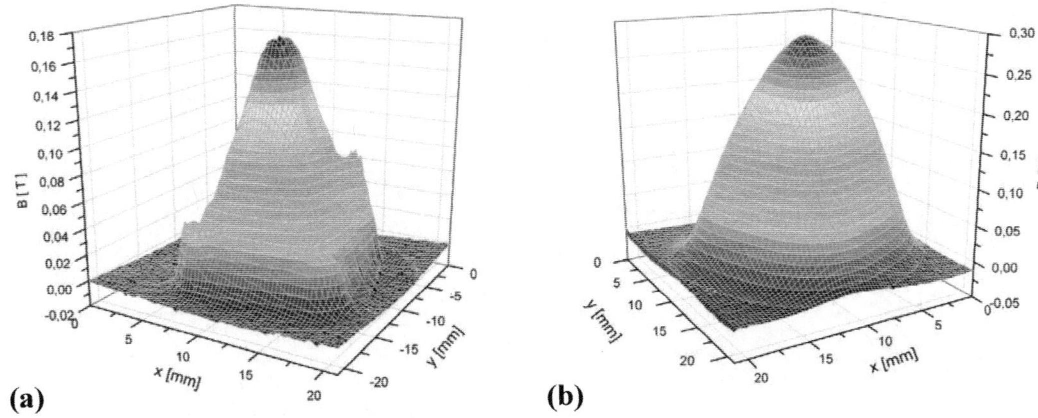

Figure 3. Measured trapped field profiles for the 20 mm diameter samples of Y123/Y211 (a) and Y123/Y_2O_3 (b) systems.

The microstructure analysis done on the polished cut of the Y123/Y211 oxygenated sample helped to explain observed irregularities in the trapped field profile (Figure 4). In polarized light, it is possible to see that the Y123/Y211 sample is not fully oxygenated. The large areas along a/b-plane are still tetragonal and nonsuperconducting (Figure 4(a)). By our opinion much higher Y211 concentration observed in the Y123/Y211 sample (Figure 4(a)) than in the Y123/Y_2O_3 sample (Figure 4(b)) is responsible for difficult oxygenation. The sample with higher Y211 concentration has higher fracture toughness [16] and therefore formation of oxygenation cracks, which are necessary for faster oxygenation [17], is difficult.

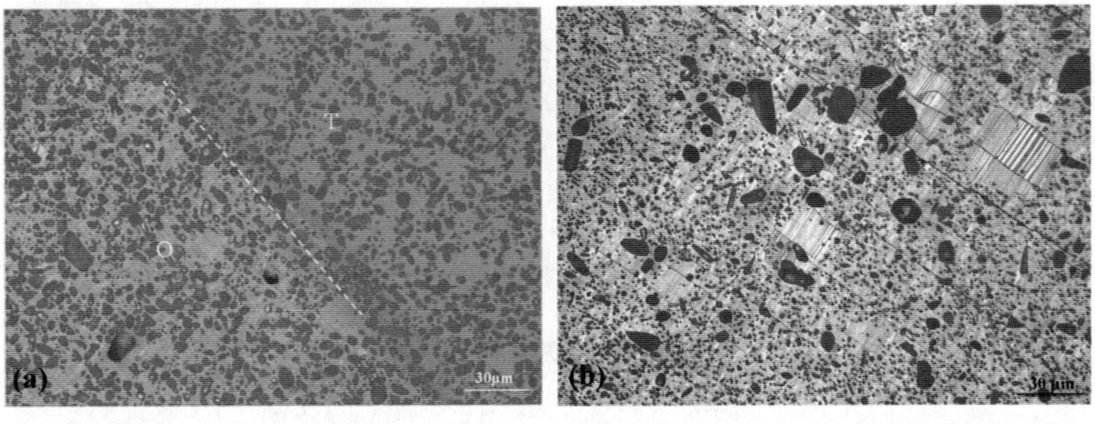

Figure 4. Microstructure after oxygenation process 400 °C/150h of the samples Y123/Y211, where we can see orthorhombic (O) and tetragonal Y123 phase (T) (a) and Y123/Y_2O_3 with orthorhombic structure in entire area of the sample (b).

The reason of higher Y211 concentration in the Y123/Y211 sample is a loss of melt at the melting temperature. We tested different substrates from the point of melt losses at different melting temperatures. The results are presented in Figure 5. The lower melt loss and change in the sample diameter were observed for the substrate in the form of ZrO_2 bars then for the Y_2O_3/Yb_2O_3 combined substrate. The melt loss was lower for lower melting temperatures. This positive influence of ZrO_2 substrate is also illustrated by observed sample microstructures (Figure 6). The size Y211 particles are similar for both tested substrates. The significant difference is in the Y211 volume fraction, which is much higher for the Y_2O_3/Yb_2O_3 combined substrate where higher melt loss was observed.

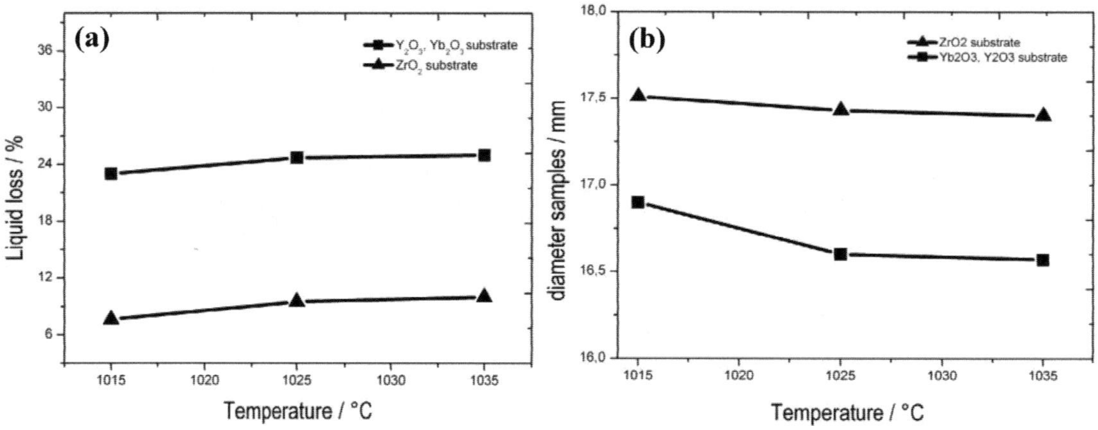

Figure 5. Melt losses (a) and changes in sample diameter (b) during growth of Y123 single-grain for different substrates (ZrO_2 and Y_2O_3/Yb_2O_3) at different T_{max} = 1015 °C, 1025 °C and 1035 °C.

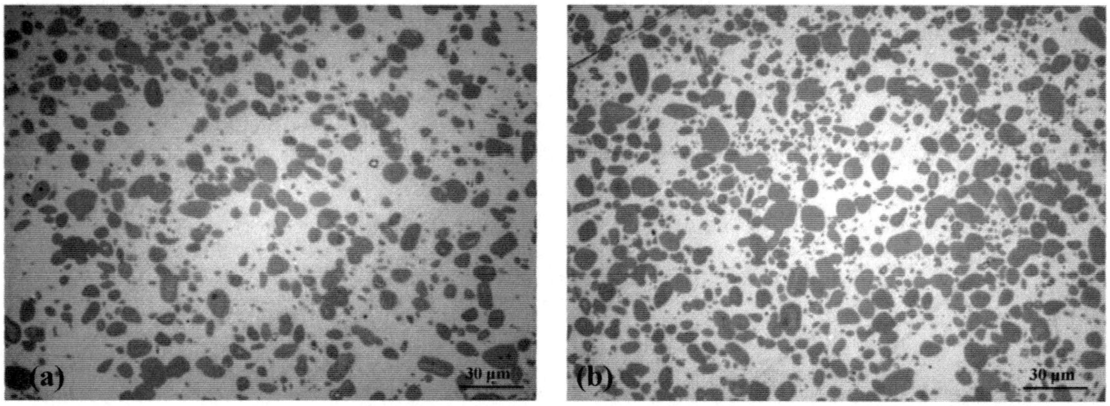

Figure 6. Microstructure of Y123/Y211 single-grain samples for different substrates and T_{max} = 1025°C. The size of Y211 particles is similar but concentration of Y211 is lower for ZrO_2 substrate (a) than for Y_2O_3/Yb_2O_3 substrate (b).

4. Conclusion

YBCO single-grain bulk superconductor with nominal composition 1 mol. Y123 + 0.25 mol. Y_2O_3 + 1wt.% CeO_2 and 70 wt.% Y123 + 30 wt.% Y211 + 1 wt.% CeO_2 were prepared by TSMG process in air. It is shown that CuO formed during the reaction of Y_2O_3 and CeO_2 with Y123 and segregates to the rest of melt what decreases the peritectic temperature of the nonsolidified part of the sample and causes growth stop in both studied systems. Moreover, high melt losses in the Y123/Y211 system with Y_2O_3/Yb_2O_3 substrate leads to the high volume fraction of Y211 particles and to difficult oxygenation. It is also shown that the melt losses are lower at lower melting temperature and for the ZrO_2 substrate.

Acknowledgment
This work was realized within the framework of the projects: Centre of Excellence of Advanced Materials with Nano- and Submicron Structure (ITMS 26220120019), Infrastructure Improving of Centre of Excellence of Advanced Materials with Nano- and Submicron Structure (ITMS 26220120035), New Materials and Technologies for Energetic (ITMS 26220220061), Research and Development of Second Generation YBCO Bulk Superconductors (ITMS 26220220041), APVV No. 0330-12, VEGA No. 2/0121/16 and Stefanik Project SK-FR-2013-0025, SAS Centre of Excellence: CFNT MVEP, PhysNet (ITMS 26110230097), NANOKOP (ITMS 26110230061).

References
[1] Zhai W, Shi Y, Durrell J H, Dennis A R, Zhang Z and Cardwell D A 2015 *Cryst. Growth Des.* **15** 907–914
[2] Yan S B, Sun J L, Li T Y, Cheng L and Yao X 2011 *Supercond. Sci. Technol.* **24** 075007
[3] Kenfaui D, Sibeud P F, Louradour E, Chaud X and Noudem J G 2014 *Advanced Functional Materials* **24** 3996–4004
[4] Plechacek V, Jirsa M, Rameš M, Muralidhar M 2012 *Physics Procedia* **36** 538
[5] Chen Y Y, Cui X X, Yao X 2014 *Progress in Materials Science* **1** 68
[6] Sawh R P, Weinstein R, Carpenter K, Parks D, Davey K 2013 *Superconductor Science and Technology* **26** 105014
[7] Diko P, Šefčiková M, Zmorayová K, Kalmanová M, Volochová D, Piovarči S 2012 *Journal of Crystal Growth* **338** 239–243
[8] Volochova D, Diko P, Radusovska M, Antal V, Piovarci S, Zmorayova K, Sefcikova M 2012 *Journal of Crystal Growth* **353** 31–34
[9] Kim Ch J, Kim K B, Park H W, Sung T H, Kuk I H, Hong G W 1996 *Supercond. Sci. Technol.* **9** 76
[10] Kim Ch J, Hong G W 1999 *Supercond. Sci. Technol.* **12** R27
[11] Kim Ch J, Jee Y A and Hong G W 2000 *Supercond. Sci. Technol.* **13** 709
[12] Zhai W, Shi Y, Durrell J H, Dennis A R, Cardwell D A, 2014 *Cryst. Growth Des.* **14** 6367- 6375
[13] Thoma M, Shi Y, Dennis T, Dùrrell J, Cardwell D 2015 *Journal of Crystal Growth* **412** 31-39
[14] Volochova D, Diko P, Antal V, Radusovska M, Piovarci S 2012 *Journal of Crystal Growth* **356** 75–80
[15] Volochová D, Diko P, Radušovská M, Piovarči S, Antal V, Zmorayová K, Šefčiková M 2013 *J Supercond Nov Magn* **26** 885–889
[16] Sakai N, Munakata F, Diko P, Takebayashi S, Yoo S I, Murakami M 1997 *Advances in Superconductivity X* vol. 1 645-648
[17] Diko P 2004 *Supercond. Sci. Technol.* **17** R45-R58

Melting of $SmBa_2Cu_3O_{7-y}$-seeds during preparation of YBCO Bulk Superconductors by Infiltration Growth Method

L Vojtkova, P Diko and D Volochová

Institute of Experimental Physics, Slovak Academy of Sciences, Watsonova 47, 04001 Košice, Slovakia

E-mail: vojtkova@saske.sk , dikos@saske.sk , volochova@saske.sk

Abstract. The conditions for single-grain growth of YBCO bulk superconductors by top seeded infiltration growth were tested. It is shown that the interaction of melt formed from $BaCuO_2$ + CuO + Y_2O_3 precursor with the Sm123 seed causes dissolution of the seed at maximum melting temperature 1045 °C. Experiments with low weight Y211 pellet confirmed that the low concentration of Y in the infiltration melt is responsible for this effect. The most effective way suppressing the seed dissolution was shown to be the insertion of Y123 + Y211 buffer layer between the seed and the Y211 pellet. This buffer layer possesses the melt which is saturated with yttrium what prevents dissolution of the seed.

1. Introduction

Single grain YBCO bulk superconductors are perspective and extensive studied materials for practical applications. There are several ways how to prepare YBCO bulk superconductors [1][2][3][4]. Top seeded infiltration growth method gives good assumptions for production of bulk single crystals with fine and uniformly distributed Y211 particles in Y123 matrix, low porosity and with big dimensions and also with complicated shapes [5]. Some authors [6-10] describe the various ways to prepare YBCO bulk single crystal by infiltration method. In general process consists of heating to melting temperature, infiltration of liquid source to Y211 pellet, cooling to peritectic transformation, slow cooling through interval of peritectic transformation and furnace cooling to room temperature. Temperature and time of individual steps as well as mass of particular components are very important for obtaining resultant bulk single crystal without parasitic ones. Problems associated with transport of atoms between liquid phase and the seed are described by Jee [11]. Some authors solve various problems connected with diffusion between the seed and liquid phase using buffer layer at crystal growth by TSMG process [12-17] and TSIG process [18].

In this work influence of maximum infiltration temperature and configuration of individual layers of experiment were studied.

2. Experimental materials and methods

2.1. Experimental material

As initial material $Y_2Ba_1Cu_1O_5$ (Y211) powder (Solvay Fluor GmbH, Germany) with a maximum particle size $d_{max.}$ = 5 μm was used. The 12 g of the powder was compacted by uni-axial pressing to Y211 pellet with diameter 20 mm. Liquid source pellet (LSP) was prepared as a mix of Y_2O_3 (Heraeus

Feinchemikalien und Forschungsbedarf Gmbh), CuO (Huizhoutian Yi Rare Material co., Ltd) and $BaCuO_2$ (prepared by calcination) powders in ratio 1:6:10 respectively in accordance to reaction 1.

$$Y_2O_3 + 6\ CuO + 10\ BaCuO_2 \rightarrow Y_2BaCuO_5 + 3\ Ba_3Cu_5O_8 \quad (1)$$

Powders were mixed in grinding mill with ZrO_2 bowl for 20 minutes. Mixed powders were uni-axially pressed into 20 gram pellet with diameter 30 mm. The diameter of liquid source is bigger than Y211 pellet for ensuring of complete infiltration and better support for final YBCO sample.

2.2. Growth experiments

Growth experiments of samples were inspired by work of Guo-Zheng Li and others [10]. The samples were heated to melting temperature $T_m = 1045°C$, heating rate 100°C/h, holding time at T_m 3 hours. Then the samples were cooled to a temperature $T_p = 998°C$ by rate 60°C/h and from 998°C to 982°C were slowly cooled for 64 hours (rate 0.25 °C/h).

The time-temperature regime is presented in Figure 1. In order to find optimal regime of YBCO single crystal growth from the seed, temperatures of solidification growth (T_p and T_o from Figure 1) were varied approximately 5°C up and down.

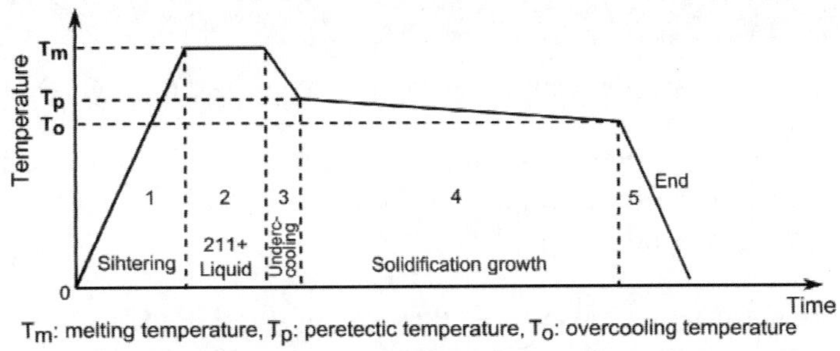

Figure 1. Time-temperature regime of single crystal Y123 production by infiltration growth method

Macrostructure observation and microstructure analysis were performed on polished specimens of the processed samples using an optical microscope equipped with a polarizer. EDX analysis was also performed.

3. Results and discussion

Some experiments were performed with samples composed of layers with arrangement as shown in Figure 2. Bottom layer was formed of MgO single crystal. On the MgO single crystal Yb_2O_3 pellet was placed. These bottom layers was used in order to avoiding reaction of liquid source pellet with Al_2O_3 base. The upper part consisted of Y211 pellet and Sm123 single crystal, prepared by TSMG process, as a seed placed on that.

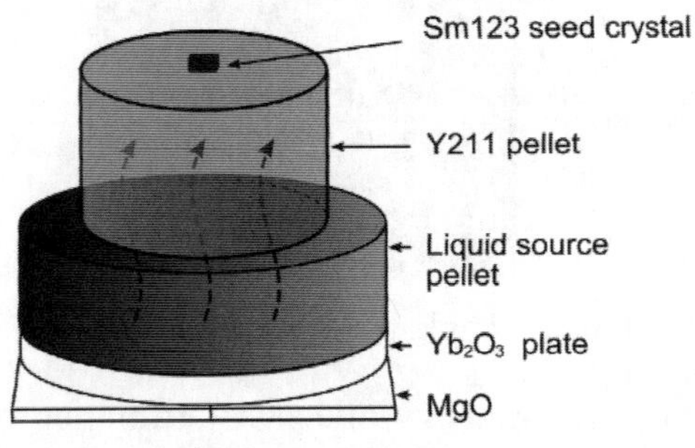

Figure 2. Schematic arrangement of individual layers of YBCO single crystal production by infiltration growth method

We have found that the seed melting occurs at heating to melting temperature T_m = 1045°C. In Figure 3 microstructures of three samples sectioned along the seed in the c-axis direction are presented. All three samples were identically heat treated, but seeds with different height were used. In Figure 3a the microstructure of sample with biggest seed is shown. It can be seen that lower part of seed was melted and subsequently crystallized at cooling as polycrystalline material. The upper part of seed remained unchanged in single crystal form. On the other hand in Figure 3c the sample with smallest seed is presented. In this case the seed was completely melted and mixed with other parts of the sample. The middle sample (Figure 3b) was treated by using the seed with intermediate height. The whole seed was melted and crystallized in polycrystalline form at cooling.

Figure 3. Microscopic analysis of samples sectioned through seed in axis c.

It is reasonable to suppose that the reason for the seed melting is diffusion of Sm from the seed to the sample melt which starts at the seed/sample interface. This process leads to decomposition of Sm123 phase according to reaction (S-solid, L-liquid, G-gas):

$$2(SmBa_2Cu_3O_x) \rightarrow (3BaO\cdot 5CuO)_{L(Sm)} + Sm_2BaCuO_5 \qquad (2)$$

The kinetic of this reaction is controlling destruction of the seed. It is faster at higher temperatures due to faster Sm diffusion in the sample melt as well as due to higher solubility of Sm in the melt at higher temperatures [19]. The seed is in fact composite of Sm123 crystal and Sm211 particles, therefore it keeps its original shape even if the Sm123 phase is decomposed according to reaction (2). Then, during cooling the self nucleation of Sm123 crystals leads to polycrystalline form (Figure 3).

In order to avoid the seed dissolution we had to decrease maximum temperature of infiltration process below 1045°C. An effective maximum temperature of infiltration process was determined to be 1025°C where the single-grain YBCO sample grows from the seed (Figure 4).

Figure 4. Sample prepared with $T_{max.}$ = 1025°C

Higher melting temperature of the process influences infiltration of the melt to Y211 pellet and homogeneity of the melt [20, 21]. If melting temperature of the melt of infiltration process is low, some clusters with structure similar to solid form of Y123 can exist in the melt. These clusters can serve as seeds of Y123 crystals at cooling. If infiltration temperature is higher there are less clusters and clusters have smaller dimensions. Thus, for the growth of single crystal without parasitic crystals, higher infiltration temperature should be used. However, this leads to dissolution of Sm123 type seeds. One possibility how to eliminate degradation of the seed is to saturate RE concentration in the melt. The solubility limit of Sm in the sample melt is about 2 wt. % [19]. This possibility was tested with the small Y211 sample, as presented in Figure 5, (diameter 8 mm, weight 0,5g). The weight of the seed was enough to saturate RE concentration (Sm + Y) in the sample melt by partial dissolution of it in the melt. The sample was identically heat treated with the same holding time at maximum melting temperature 1045°C of the process. As a result of using small sample the perfect Y123 single crystal was grown from the seed.

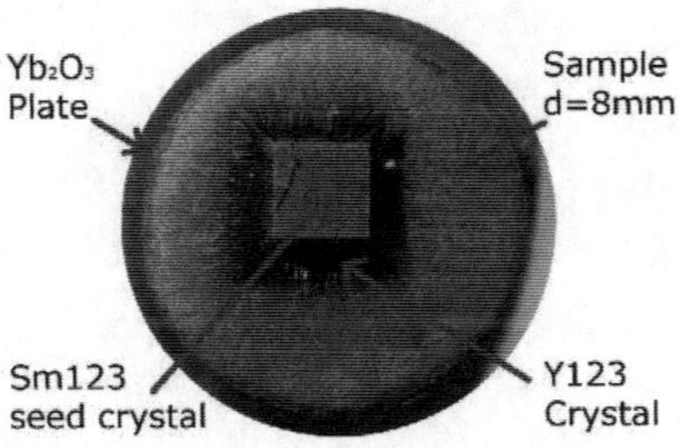

Figure 5. The change of ratio of seed size versus sample size in favour of seed

Conducted experiments and their results point on very low concentration of Y in the infiltrated melt which is obviously caused by slow dissolution of Y211 particles in the melt at the infiltration process. This is in contrast with the melt which is formed from Y123 phase at TSMG process. During peritectic decomposition of Y123 phase according to reaction (S-solid, L-liquid, G-gas):

$$(2YBa_2Cu_3O_x)_S \rightarrow (Y_2BaCuO_5)_S + (3BaO{\cdot}5CuO)_{L(Y)} + (O_2)_G \tag{3}$$

the melt L is saturated with Y and the Sm123 type seed is not destructed by reaction (3). From this consideration an idea comes to put an interlayer containing Y123 phase between the Sm123 seed and the Y211 pellet. This arrangement should provide Y reach melt near the seed and to prevent its melting.

For this experiment the same powders and pellets (Y211 and liquid source pellet) as in previous experiment were used. The main difference was in using of buffer layer between Sm123 seed and Y211 pellet. Buffer layer was prepared as a mixture of Y123 and Y211 (weight ratio 7:3) powders with addition of 1 wt. %, CeO_2 [22]. The buffer layer powders were also mixed in grinding mill and uni-axially pressed into the 0.15 gram pellet with diameter 5 mm. The arrangement of individual layers is presented in Figure 6 and consists of MgO single crystal, Yb_2O_3 plate, liquid source pellet, Y211 pellet, buffer layer (composed of Y123 and Y211) and Sm123 seed. All were placed on Al_2O_3 base.

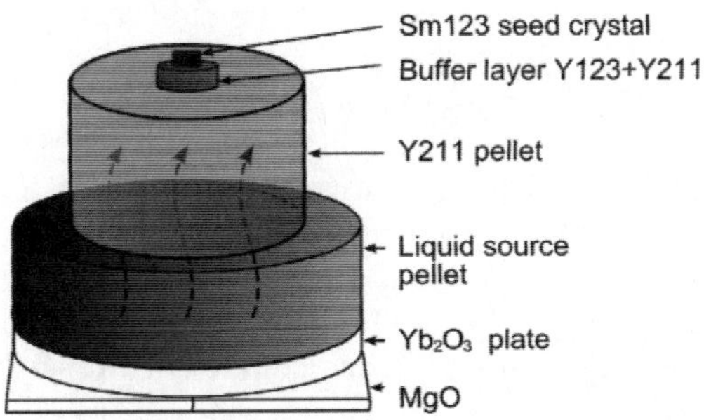

Figure 6. Schematic arrangement of individual layers of YBCO single crystal production by infiltration growth method with buffer layer

In Figure 7 the sample from experiment with used buffer layer is presented. Maximum temperature of infiltration process was 1045°C. Seed remained undissolved and grown single crystal can be seen (Figure 7a). In Figure 7b microstructure of the sample sectioned through seed and buffer layer along the a/c-plane is presented. The growth sector boundary (GSB) of Y123 single crystal progressing from seed through buffer layer to Y211 pellet can be seen.

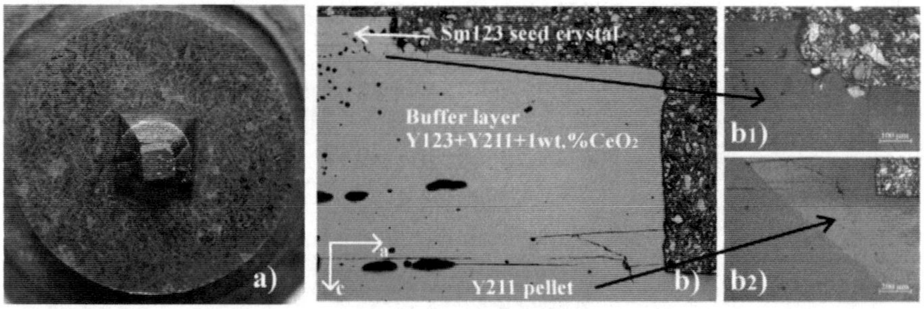

Figure 7. Sample prepared using buffer layer at T_{max} = 1045°C, a) top view of the sample, b) part of the sample cut along the a/c-plane, b_1) detail of the GSB in the buffer layer, b_2) detail of the GSB in the buffer layer and in the Y123 single crystal.

4. Conclusions

The dissolution of the Sm123 type seed at 1045 °C melting temperature was observed at top seeded infiltration growth of YBCO bulk single-grain samples. We found three possibilities how to protect the seed. The first is to lower the melting temperature. The maximum melting temperature of infiltration process was determined to 1025°C. The second possibility is to decrease the weight of the Y211 pellet. The experiments with smaller Y211 bulk pellet showed that the main factor of dissolution of the seed is saturation of the melt with the rare earth element. If the Y211 pellet of 0.5 gram in weight was used, the single-grain sample has been grown from the Sm123 seed also when maximum melting temperature 1045 °C was used. The insertion of Y123+Y211 buffer layer between the seed and Y211 pellet has been shown to be the third solution of the problem. The buffer layer possesses the melt which is saturated with yttrium what prevents dissolution of Sm from the seed in this melt even at melting temperature 1045 °C.

Acknowledgements

This work was realized within the framework of the projects: ITMS 26220120019, ITMS 26220120035, ITMS 26220220061, ITMS 26220220041, SAS Centre of Excellence: CFNT MVEP, PhysNet ITMS 26110230097, NANOKOP ITMS 26110230061, APVV No. 0330-12, VEGA No. 2/0121/16 and Stefanik Project SK-FR-2013-0025.

References

[1] Cardwell D 2003 *Handbook of Superconducting Materials.* (British: Institute of Physics Publishing)
[2] Maruyama O, Ohkuma T, Masuda T, Ohya M, Mukoyama S, Yagi M, Saitoh T, Aoki N, Amemiya N, Ishiyama A and Hayakawa N 2012 Development of REBCO HTS Power Cables *Physics Procedia* **36** 1153
[3] Haindl S, Eisterer M, Horhager N, Weber H W, Walter H, Shlyk L, Krabbes G, Hari Babu N and Cardwell D A 2005 Novel methods to characterize bulk RE-BCO superconductors. *Physica C* **426–431** 625
[4] Noudem J G, Meslin S, Horvath D, Harnois C, Chateigner D, Eve S, Gomina M, Chaud X and Murakami M 2007 Fabrication of textured YBCO bulks with artificial holes *Physica C* **463–**

465 301
[5] Umakoshi S, Ikeda Y, Wongsatanawarid A, Kim C-J and Murakami M 2011 Top-seeded infiltration growth of Y-Ba-Cu-o bulk superconductors *Physica C* **471** 843
[6] Guo-Zheng Li and Wan-Min Yang 2011 Fabrication of single-grain $GdBa_2Cu_3O_{7-x}$ bulk superconductors with a new kind of liquid source by the top seeded infiltration and growth technique *Materials Chemistry and Physics* **129** 288
[7] Fang H and Ravi-Chandar K 2000 Fabrication of Y123disk by the seeded infiltration and growth method *Physica C* **340** 261
[8] Pavan Kumar Naik S 2012 Effect of infiltration temperature on the properties of infiltration growth processed YBCO superconductor *Journal of Alloys and Compounds* **551** 318
[9] Devendra Kumar N 2013 Preform optimization in infiltration growth process: An efficient method to improve the superconducting properties of $YBa_2Cu_3O_{7\_d}$ *Physica C* **495** 55
[10] Guo-Zheng Li, De-Jun Li, Jian-Hua Deng and Wan-Min Yang 2013 Effects of liquid source mass and slow-cooling time on the infiltration growth of single-grain YBaCuO bulk superconductors *Journal of Alloys and Compounds* **551** 318
[11] Jee Y A, Hong G-W, Kim C-J and Sung T-H 1998 Dissolution and Resolidification of $SmBa_2Cu_3O_{7-y}$ seed during top seeded melt texturing of $YBa_2Cu_3O_{7-y}$ *IEEE Transactions on Applied Superconductivity* **9(2)** 2097
[12] Kim C-J, Lee J H, Park S-D, Jun B-H, Han S C and Han Y H 2011 Y_2BaCuO_5 buffer block as a diffusion barrier for samarium in top seeded melt growth processed $YBa_2Cu_3O_{7-y}$ superconductors using a $SmBa_2Cu_3O_{7-d}$ seed. *Superconductor Science and Technology* **24(1)** 015008
[13] Li T Y, Cheng L, Yan S B, Sun L J, Yao X, Yoshida Y and Ikuta H 2010 Growth and superconductivity of REBCO bulk processed by a seed/buffer layer/precursor construction. *Superconductor Science and Technology* **23(12)** 125002
[14] Devendra Kumar N, Shi Y, Zhai W, Dennis A R, Durrell J H and Cardwell D A 2015 Buffer Pellets for High-Yield, Top-Seeded Melt Growth of Large Grain Y–Ba–Cu–O Superconductors. *Crystal Growth & Design* **15(3)** 1472
[15] Shi Y, Kumar Namburi D, Zhao W, Durrell J H, Dennis A R and Cardwell D A 2016 The use of buffer pellets to pseudo hot seed (RE)–Ba–Cu–O–(Ag) single grain bulk superconductors. *Superconductor Science and Technology* **29(1)** 015010
[16] Shi Y-H, Dennis A R and Cardwell D A 2015 A new seeding technique for the reliable fabrication of large, SmBCO single grains containing silver using top seeded melt growth. *Superconductor Science and Technology* **28(3)** 035014
[17] Xu H H, Chen Y Y, Cheng L, Yan S B, Yu D J and Yao X 2012 YBCO-buffered NdBCO film with higher thermal stability in seeding REBCO growth. *Superconductor Science and Technology* **25(3)** 035014
[18] Namburi D K, Shi Y, Palmer K G, Dennis A R, Durrell J H and Cardwell D A 2015 An improved top seeded infiltration growth method for the fabrication of Y–Ba–Cu–O bulk superconductors. *Journal of the European Ceramic Society* **36** 615
[19] Krabbes G, Fuchs G, Canders W-R, May H and Palka R 2006 *High Temperature Superconductor Bulk Materials* (Weinheim: WILEY-VCH)
[20] Xu H H, Cheng L, Yan S B, Yu D J, Guo L S and Yao X 2012 Recycling failed bulk YBCO superconductors using the NdBCO/YBCO/MgO film-seeded top-seeded melt growth method *Journal of applied physics* **111** 103910
[21] Yan S B, Sun L J, Li T Y, Cheng L and Yao X 2011 Differences in the thermal stability of $REBa_2Cu_3O_{7-x}$ (RE = Y, Nd) thin films investigated by high temperature in situ observation and melt-texture growth *Supercond. Sci.Technol.* **24** 075007
[22] Volochová D, Diko P, Antal V, Radušovská M and Piovarči S 2012 Influence of Y_2O_3 and CeO_2 addition on growth of YBCO bulk superconductors *Journal of Crystal Growth* **356** 75

Institute of Physics
Dirac House, Temple Back
Bristol BS1 6BE UK

ISSN: 1742-6588
ISBN 978-1-5108-2237-5